3D 图形系统设计与实现

[美] 乔纳斯·戈麦斯（Jonas Gomes）
[美] 路易斯·维霍（Luiz Velho）　　　　著
[美] 马里奥·科斯塔·苏萨（Mario Costa Sousa）

周建娟　译

清华大学出版社
北　京

内 容 简 介

本书详细阐述了与3D图形系统设计与实现相关的基本解决方案，主要包括对象和图形设备、交互式图形界面、几何体、颜色、数字图像、3D场景描述、三维几何体模型、建模技术、层次结构和体系结构对象、视见相机转换、视见的表面剪裁、光栅化、可见表面计算、局部光照模型、全局光照、贴图技术、着色机制，以及三维图形系统等内容。此外，本书还提供了相应的示例，以帮助读者进一步理解相关方案的实现过程。

本书适合作为高等院校计算机及相关专业的教材和教学参考书，也可作为相关开发人员的自学教材和参考手册。

Design and Implementation of 3D Graphics Systems 1st Edition/by Jonas Gomes, Luiz Velho and Mario Costa Sousa/ISBN:978-1-4665-7121-1

Copyright © 2013 by CRC Press.

Authorized translation from English language edition published by CRC Press, part of Taylor & Francis Group LLC; All rights reserved;

本书原版由Taylor & Francis出版集团旗下CRC出版公司出版，并经其授权翻译出版。版权所有，侵权必究。

Tsinghua University Press is authorized to publish and distribute exclusively the **Chinese(Simplified Characters)** language edition.This edition is authorized for sale throughout **Mailand of China**.No part of the publication may be reproduced or distributed by any means, or stored in a database or retrieval system, without the prior written permission of the publisher.

本书中文简体翻译版授权由清华大学出版社独家出版并限在中国大陆地区销售。未经出版者书面许可，不得以任何方式复制或发行本书的任何部分。

Copies of this book sold without a Taylor & Francis sticker on the cover are unauthorized and illegal.

本书封面贴有Taylor & Francis公司防伪标签，无标签者不得销售。

北京市版权局著作权合同登记号 图字：01-2014-0802

版权所有，侵权必究。侵权举报电话：010-62782989 13701121933

图书在版编目（CIP）数据

3D图形系统设计与实现 ／（美）乔纳斯·戈麦斯（Jonas Gomes），（美）路易斯·维霍（Luiz Velho），（美）马里奥·科斯塔·苏萨（Mario Costa Sousa）著；周建娟译． —北京：清华大学出版社，2020.1

书名原文：Design and Implementation of 3D Graphics Systems

ISBN 978-7-302-54447-0

Ⅰ. ①3… Ⅱ. ①乔… ②路… ③马… ④周… Ⅲ. ①三维动画软件 Ⅳ. ①TP391.414

中国版本图书馆CIP数据核字（2019）第265431号

责任编辑：贾小红
封面设计：刘 超
版式设计：文森时代
责任校对：马军令
责任印制：沈 露

出版发行：清华大学出版社
网　　址：http://www.tup.com.cn, http://www.wqbook.com
地　　址：北京清华大学学研大厦A座
邮　　编：100084
社 总 机：010-62770175
邮　　购：010-62786544
投稿与读者服务：010-62776969, c-service@tup.tsinghua.edu.cn
质量反馈：010-62772015, zhiliang@tup.tsinghua.edu.cn

印 装 者：清华大学印刷厂
经　　销：全国新华书店
开　　本：185mm×230mm
印　　张：22.5
字　　数：448千字
版　　次：2020年1月第1版
印　　次：2020年1月第1次印刷
定　　价：119.00元

产品编号：055781-01

译 者 序

　　3D 计算机图形学在科学可视化、CAD/CAM、电影工业以及游戏娱乐业方面均有所应用。本书介绍了计算机图形学中实践操作方面的内容，相关内容主要集中在该领域的基本算法、实现问题，以及图形系统中各个组件之间的关系。具体来说，对应于图形系统的主架构选择方案，本书将开发 3 个完整的图形系统。

　　本书以一种基本的方式阐述计算机图形学，并考查这门学科的基本原理，目的是为以后的学习打下坚实的基础。然而，本书并非是一本纯理论书籍，除了对相关内容进行了全面、系统的讲解以外，其设计思想、数据结构和算法均辅以对应的代码示例，以帮助读者进一步理解计算方案的实现过程。

　　在本书的翻译过程中，除周建娟外，王辉、刘晓雪、张博、张华臻、刘璋、刘祎等人也参与了部分翻译工作，在此一并表示感谢。

译　者

前　　言

本书涵盖了几何建模和渲染三维场景过程中计算方面的内容，特别是 3D 图形系统体系结构方面的知识。本书讨论了基本的 3D 计算机图形学算法，并全部采用 C 语言加以实现。为了进一步构建图形系统，本书还补充了相关的库例程。读者可访问 http://www.crcpress.com/product/isbn/9781568815800 下载相关的库例程和其他补充材料。

本书的姊妹篇 Computer Graphics: Theory and Practice [Gomes et al. 12]则重点讲述了计算机图形学中的概念、基础理论和模型、计算应用数学的抽象范例，并用于解决计算机图形学不同领域的问题。

本书是多位同仁联袂合作的结果。本书的出版要归功于 Paulo Roma Cavalcanti，他不仅教授了这门课程，而且还提供了课程笔记；同时，他也是本书最初的审校者。Luiz Henrique de Figueiredo 对本书的部分章节进行了详细的审读，并制作了本书中的一些插图。此外，还要感谢 Margareth Prevot（IMPA，VisGraf Lab），他参与了本书中图像的制作。另外，我们也要感谢每一位为我们提供帮助的人。

除此之外，其他同仁也审读了本书的首稿，并提出了宝贵的意见，在此鸣谢 Antonio Elias Fabris、Romildo José da Silva、Cícero Cavalcanti、Moacyr A. Silva、Fernando W.da Silva、Marcos V. Rayol Sobreiro、Silvio Levy 和 Emilio Vital Brazil，我们衷心地感谢大家。同时还要感谢 Jamie McInnis、Sarah Chow 和 Patricia Rebolo Medici 对本书的意见和建议，以及他们对本书的精心编辑和校对。

本书内容得益于 IMPA（VisGraf Lab）计算机图形学实验室、计算机科学系、Calgary 大学交互式建模和可视化小组（插图/iRMV）/计算机图形学研究实验室卓有成效的教学成果和研究环境。这里，我们衷心感谢全体成员对我们的支持。最后，还要感谢 NSERC/AITF/Foundation CMG Industrial Research Chair 项目对我们的大力帮助。

目 录

第 1 章 概述 1
   1.1 计算机图形学 1
   1.2 应用领域和应用程序 1
   1.3 研究方法 2
   1.4 系统架构 2
   1.5 实现和扩展 3
   1.6 实现范例 3
   1.7 图形标准 4
   1.8 高级应用程序和后续发展 4
   1.9 本书内容 5
   1.10 补充材料 6

第 2 章 对象和图形设备 7
   2.1 图形对象 7
      2.1.1 图形对象的描述 8
      2.1.2 图形对象的离散化和重构 8
   2.2 图形设备和表达 10
      2.2.1 向量设备 10
      2.2.2 光栅化（矩阵）设备 10
   2.3 图形设备分类 11
      2.3.1 概念 12
      2.3.2 分类 12
   2.4 图形工作站 13
      2.4.1 窗口系统 13
      2.4.2 视图转换 14
   2.5 GP 图形包 15
      2.5.1 GP 特征 15
      2.5.2 GP 中的颜色属性 16

    2.5.3 GP 中对象的数据结构 ... 17
    2.5.4 控制例程 ... 18
    2.5.5 视见例程 ... 19
    2.5.6 绘制例程 ... 21
    2.5.7 图形输入和交互例程 ... 22
  2.6 补充材料 ... 24
  2.7 本章练习 ... 25

第 3 章 交互式图形界面 ... 27
  3.1 创建交互式程序 ... 27
  3.2 交互基础 ... 27
    3.2.1 图形反馈 ... 28
    3.2.2 逻辑输入元素 ... 28
    3.2.3 概览 ... 28
  3.3 界面机制 ... 29
    3.3.1 非交互式 ... 29
    3.3.2 事件驱动 ... 29
    3.3.3 回调模型 ... 30
    3.3.4 包含多个视图的回调 ... 31
  3.4 界面对象 ... 32
    3.4.1 多视口 ... 32
    3.4.2 基于视图的回调 ... 36
  3.5 工具箱 ... 40
    3.5.1 基本元素 ... 40
    3.5.2 tk 包 ... 41
    3.5.3 示例 ... 46
  3.6 多边形直线编辑器 ... 47
  3.7 回顾 ... 54
  3.8 补充材料 ... 55
  3.9 本章练习 ... 56

第 4 章 几何体 ... 57
  4.1 计算机图形学中的几何体 ... 57

4.1.1 应用和功能 57
4.1.2 计算内容 57
4.1.3 方案汇总 57
4.2 欧几里得空间 58
4.2.1 定义 58
4.2.2 元素和操作 58
4.2.3 度量属性 60
4.2.4 坐标和基 61
4.3 欧几里得空间中的转换 62
4.3.1 线性转换 62
4.3.2 等距 63
4.3.3 仿射转换 63
4.4 投影空间 63
4.4.1 投影空间模型 64
4.4.2 标准化和齐次坐标 64
4.4.3 齐次表达 65
4.5 $\mathbb{RP}^3$ 中的投影转换 66
4.6 几何体对象的转换 73
4.6.1 转换操作修正 73
4.6.2 转换点和方向 73
4.6.3 转换射线 74
4.6.4 切平面上的转换 76
4.6.5 转换的双重解释 76
4.7 补充材料 77
4.7.1 小结 77
4.7.2 程序设计层 78
4.8 本章练习 78

第 5 章 颜色 81
5.1 颜色的基本知识 81
5.1.1 颜色的波长模型 81
5.1.2 物理颜色系统 82

- 5.1.3 色彩的心理学研究 82
- 5.1.4 颜色计算 84
- 5.2 设备颜色系统 85
  - 5.2.1 颜色的处理 85
  - 5.2.2 RGB-CMY 转换 85
- 5.3 颜色规范系统 87
  - 5.3.1 亮度：色度分解 87
  - 5.3.2 颜色选择的 HSV 系统 88
- 5.4 离散化颜色实体 92
- 5.5 补充材料 93
  - 5.5.1 资料链接 93
  - 5.5.2 回顾 94
- 5.6 本章练习 94

第 6 章 数字图像 95
- 6.1 基础知识 95
  - 6.1.1 图像的离散和连续模型 95
  - 6.1.2 图像的量化 96
  - 6.1.3 矩阵表达 97
- 6.2 图像的表现格式 97
  - 6.2.1 数据结构 97
  - 6.2.2 访问图像矩阵 99
- 6.3 图像编码 100
  - 6.3.1 PPM 格式 100
  - 6.3.2 直接编码 100
- 6.4 补充材料 102
  - 6.4.1 修正 102
  - 6.4.2 图像格式 103
- 6.5 本章练习 103

第 7 章 3D 场景描述 105
- 7.1 三维场景 105
  - 7.1.1 三维场景的元素 105

7.1.2 三维场景表达 106
7.1.3 场景描述语言 106
7.2 语言概念 107
7.2.1 表达式语言 107
7.2.2 表达式中的语法和语义 108
7.2.3 程序的编译和解释 109
7.2.4 语言开发工具 110
7.3 扩展语言 110
7.3.1 语法分析器 110
7.3.2 词法分析器 112
7.3.3 符号分析器 115
7.3.4 参数和值 117
7.3.5 节点和表达式 119
7.3.6 辅助函数 121
7.4 子语言和应用程序 123
7.4.1 基于扩展语言的接口 123
7.4.2 实现语义 124
7.4.3 生成解释器 125
7.5 补充材料 125
7.5.1 修正 125
7.5.2 扩展 126
7.5.3 相关信息 126
7.6 本章练习 127

第8章 三维几何体模型 129
8.1 建模基础知识 129
8.1.1 模型和几何体描述 129
8.1.2 表达模式 131
8.2 几何图元 132
8.2.1 图元对象定义 133
8.2.2 泛型接口 134
8.2.3 图元示例 138

8.3 表面和多边形网格的近似计算 147
　　8.3.1 近似方法 147
　　8.3.2 分段式线性近似方法 147
8.4 多边形表面 147
　　8.4.1 n 边多边形 148
　　8.4.2 三角形 150
　　8.4.3 三角形列表 154
8.5 补充材料 156
8.6 本章练习 157

第9章 建模技术 159
9.1 建模系统的基础知识 159
　　9.1.1 用户界面 159
　　9.1.2 模型操作 160
　　9.1.3 建模技术 160
　　9.1.4 系统架构 160
9.2 构造模型 161
　　9.2.1 CSG 结构 162
　　9.2.2 简单的 CSG 表达式语言 164
　　9.2.3 三维场景描述语言中的 CSG 表达 166
　　9.2.4 三维场景描述语言中的 CSG 对象的解释 167
9.3 生成式建模技术 168
　　9.3.1 生成式模型的多边形近似表达 169
　　9.3.2 生成式模型的类型 171
　　9.3.3 旋转曲面 172
9.4 补充材料 172
9.5 本章练习 173

第10章 层次结构和体系结构对象 175
10.1 几何链接 175
　　10.1.1 层次结构 175
　　10.1.2 几何转换 176
　　10.1.3 仿射不变性 177

10.2 层次结构和转换 ... 178
10.2.1 栈操作 ... 178
10.2.2 转换 ... 180
10.3 对象分组 ... 183
10.3.1 层次结构描述 ... 183
10.3.2 对象 ... 183
10.3.3 分组和对象列表 ... 185
10.3.4 对象转换 ... 187
10.3.5 收集列表中的对象 ... 188
10.3.6 参数化链接 ... 189
10.4 动画 ... 191
10.4.1 动画时钟 ... 191
10.4.2 过程式动画的构建 ... 193
10.4.3 动画的执行过程 ... 193
10.5 补充材料 ... 195
10.6 本章练习 ... 196

第11章 视见相机转换 ... 199
11.1 视见处理过程 ... 199
11.1.1 视见操作和参考空间 ... 199
11.1.2 虚拟相机和视见参数 ... 200
11.1.3 定义视见参数 ... 202
11.2 视见转换 ... 205
11.2.1 相机转换 ... 206
11.2.2 剪裁转换 ... 208
11.2.3 透视转换 ... 209
11.2.4 设备转换 ... 212
11.2.5 转换序列 ... 213
11.3 视见规范 ... 214
11.3.1 初始化 ... 214
11.3.2 相机 ... 215
11.3.3 透视 ... 216

11.3.4　设备 217
　　　11.3.5　定义三维场景描述语言中的视见机制 218
　11.4　补充材料 218
　11.5　本章练习 219

第12章　视见的表面剪裁 221
　12.1　剪裁操作的基本知识 221
　　　12.1.1　空间剪裁 221
　　　12.1.2　剪裁和视见 221
　12.2　剪裁简单情形 222
　　　12.2.1　简单拒绝 222
　　　12.2.2　简单接受 223
　　　12.2.3　包含相反方向的面元 223
　12.3　两步剪裁 224
　12.4　序列剪裁 228
　12.5　补充材料 232
　12.6　本章练习 233

第13章　光栅化 235
　13.1　光栅化基础知识 235
　13.2　光栅化方法的分类 236
　13.3　渐增式方法 236
　　　13.3.1　内在型渐增式光栅化 236
　　　13.3.2　外在型渐增式光栅化 240
　13.4　基于细分的光栅化 240
　　　13.4.1　内在型细分 241
　　　13.4.2　外在型细分 242
　13.5　补充材料 244
　13.6　本章练习 244

第14章　可见表面计算 247
　14.1　基础知识 247
　　　14.1.1　场景属性和一致性 247

- 14.1.2 表达和坐标系 248
- 14.1.3 分类 248
- 14.2 Z-缓冲区 249
- 14.3 光线跟踪 251
  - 14.3.1 与三维场景对象的交点 251
  - 14.3.2 与 CSG 模型间的交点 252
- 14.4 Painter 算法 254
  - 14.4.1 近似 Z-排序 254
  - 14.4.2 完全 Z-排序 255
- 14.5 其他可见性方法 256
  - 14.5.1 空间细分 256
  - 14.5.2 递归细分 257
- 14.6 补充材料 258
- 14.7 本章练习 259

第 15 章 局部光照模型 261

- 15.1 基础知识 261
  - 15.1.1 光照 261
  - 15.1.2 光线传播 262
  - 15.1.3 表面和材质 262
  - 15.1.4 局部光照模型 263
- 15.2 光源 265
  - 15.2.1 光线传输 265
  - 15.2.2 光源的表达 267
- 15.3 局部光照 268
  - 15.3.1 光照上下文 269
  - 15.3.2 光照函数 269
- 15.4 材质 271
  - 15.4.1 描述材质 271
  - 15.4.2 材质类型 271
- 15.5 语言规范 272
- 15.6 补充材料 273

15.7 本章练习 ... 274

第16章 全局光照 ... 275
16.1 光照模型 ... 275
16.1.1 传输过程 ... 276
16.1.2 边界条件 ... 276
16.1.3 辐射度方程 ... 277
16.1.4 数值近似 ... 278
16.1.5 光照计算方法 ... 279
16.2 光线跟踪方法 ... 279
16.3 辐射度方法 ... 285
16.4 补充材料 ... 293
16.5 本章练习 ... 293

第17章 贴图技术 ... 295
17.1 基础知识 ... 295
17.1.1 贴图的概念 ... 295
17.1.2 贴图类型 ... 296
17.1.3 贴图应用 ... 296
17.2 纹理函数 ... 297
17.2.1 表达方式 ... 297
17.2.2 图像定义 ... 297
17.2.3 过程式定义 ... 299
17.3 纹理贴图 ... 300
17.4 凹凸贴图 ... 302
17.5 反射贴图 ... 304
17.6 光源贴图 ... 306
17.7 补充材料 ... 308

第18章 着色机制 ... 309
18.1 着色函数采样和重构 ... 309
18.2 采样方法 ... 309
18.3 基本的重构方法 ... 310

18.3.1 Bouknight 着色 ... 310
18.3.2 Gouraud 方法 ... 310
18.3.3 Phong 方法 ... 312
18.4 纹理属性的重构 ... 313
18.4.1 插值和投影转换 ... 313
18.4.2 纹理的有理线性插值 ... 315
18.5 图像化 ... 317
18.6 补充材料 ... 318

第 19 章 三维图形系统 ... 321
19.1 系统 A ... 321
19.1.1 生成模型 ... 321
19.1.2 基于 Z-缓冲区的渲染机制 ... 323
19.2 系统 B ... 325
19.2.1 CSG 建模机制 ... 325
19.2.2 基于光线跟踪的渲染机制 ... 326
19.3 系统 C ... 329
19.3.1 基于图元层次结构的建模 ... 329
19.3.2 基于 Painter 方法的渲染机制 ... 329
19.4 项目 ... 332
19.4.1 渲染图像的程序 ... 333
19.4.2 建模系统 ... 333
19.4.3 渲染系统 ... 335

参考文献 ... 337

第 1 章 概 述

本书介绍了计算机图形学中实践操作方面的内容,相关内容主要集中在该领域的基本算法、实现问题,以及图形系统中各个组件之间的关系。

1.1 计算机图形学

计算机图形学是一门设计几何模型和数字图像的计算处理学科。在计算机图形学中,数据和处理间的关系如图 1.1 所示。

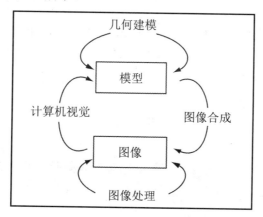

图 1.1 计算机图形学中的数据和处理

从数字图像至几何模型间的逆向转换则称作图像处理,常用于计算机图形学领域或可视化领域。本书重点讨论几何建模机制和图像合成,同时也会涉及某些图像处理方面的内容,这对于大多数图像处理类型来说还是必要的。另外,本书还将讨论图像的编码和量化操作,但其他一些较为重要的图像处理技术,如采样、重构和变型(用于纹理映射)则不在本书讨论范围内。

1.2 应用领域和应用程序

特别地,本书将深入讨论 3D 计算机图形学,同时处理 3D 场景的建模和图像合成问题。这可能是计算机图形学中最为复杂和重要的部分,且包含了较多的应用程序,如下所示。

- 科学可视化。在该领域内,计算机图形学用于可视化模拟,以及科学学科中复杂的结构,例如数学、医药和生物学。
- CAD/CAM。在该领域内,计算机图形学用于生成特效,如电影和电视中的交互式操作,以及电影公园和游戏中的娱乐应用。
- 娱乐业。在这一领域,计算机图形学用于生成电影和电视中的特效和交互式程序,以及游戏中的应用程序。

除此之外,在 2D 计算机图形学和图形界面中,还存在一些几何建模和图形合成应用。它们十分重要,并可与 3D 计算机图形学进行交互,例如,针对图像可视化和 2D 元素的构建(如用于 3D 对象建模中的平面曲线),2D 计算机图形学与图形设备关系紧密。尽管这些应用程序很重要,但鉴于其特殊性,建议单独对此予以研究,因此不在本书所讨论的范围内。

人类与计算机间的交互行为也属于计算系统中的一部分内容,在图形系统中,界面的设计往往也十分重要。本书将对此做最低限度的介绍,并假设读者已经对现有的通用窗口系统交互资源有所了解。

1.3 研究方法

本书采用了一种极简主义方法,这意味着,在理解计算机图形处理过程中,所有内容将被设置在最低限度。该方法的目的在于揭示图形系统的复杂性,以及计算机图形学中的基本处理过程。

总体来讲,全部内容覆盖了 3D 图形系统中的基本技术。针对每个问题,我们将探讨可能的解决方案及其优缺点,并以算法形式予以体现。最后,针对简单图形系统的构建,本书还将展示相关技术的具体实现过程。

1.4 系统架构

在系统层面,处理过程位于图形系统的 3 个基本模块中,进而实现了建模、图像合成和成像功能。

其中,建模模块根据场景中的对象规范创建几何表达形式。图像合成模块场景的几何描述转换为"虚拟图像",即着色功能。成像模块则根据着色功能生成数字图像。

整合上述处理过程后，将得到基本的图形系统结构，如图 1.2 所示。本书将讨论如何实现此类模块。

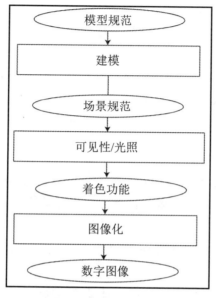

图 1.2　图形系统结构

1.5　实现和扩展

对应于图形系统的主架构选择方案，本书将开发 3 个完整的图形系统。

如前所述，出于教学目的，本书所展示系统的代码简单明了，且未过多地考虑性能和完整性问题。尽管如此，该系统依然涵盖了真实图形系统的必要功能，并可用作完整系统的雏形。对此，需要进一步提升其效率，这需要对书中的算法、功能进行局部修改，进而引入新的增益性算法。

通过考查现有方案，并在每章练习中提供相关建议，本书还将尝试某些扩展性工作。

1.6　实现范例

本书将遵循 UNIX GNU 系统开发环境中的实现范例。除了 C 语言之外，还将针对系统编译使用 make 程序，并在生成解释器时使用 yacc。该环境可以在 Windows 系统中使用 GNU 开发的公共域实用程序复制。

另外,程序间的通信将使用 C 语言中的标准输入和输出机制(stdin 和 stdout),进而查看适用于系统架构的 UNIX 管道资源。

1.7　图形标准

出于简单性考虑,本书将不使用除 GNU 之外的任何其他编程资源。这意味着,对于本书所涉及的系统开发,我们将不采用特定的库或图形标准。图形标准的重要性不言而喻,但此类标准往往更适用于开发高级程序。但本书将不使用外部编程资源,以使相关内容具有自包含性和独立性。

本书所讨论的方法和算法构成了图形标准的主体内容,如 OpenGL、VRML 语言和 Renderman。因此,深入研究这一部分内容将有助于理解和掌握相关标准。在每一章的结尾,我们还将着重强调所学算法与现有标准间的适配方式。

1.8　高级应用程序和后续发展

如前所述,本书以一种基本的方式阐述计算机图形学,并考查这门学科的基本原理,目的是为以后的学习打下坚实的基础。根据这一核心精神,未来的研究可以朝几个方向发展。考虑到读者的背景不一而同,一些专业性较强的内容将置于具体的上下文环境中,并将其与所研究的对象联系起来。

对于更加高级的研究方向,计算机图形学主要构建于至少 4 个领域上,即交互式图形系统、图形数据库、相片级成像和物理光照模拟。图 1.3 显示了此类领域和 3D 图形学之间的关系。

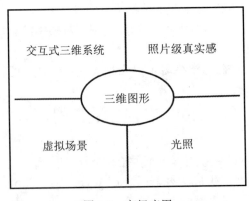

图 1.3　高级应用

在交互式图形系统领域中，OpenGL 和 Inventor 标准较为重要。在多媒体和分布式图形应用程序领域中，VRML 语言则占主导地位。在照片级成像领域中，则有 Renderman 标准。在物理光照模拟领域中，则需要重点考查与辐射度相关的内容。

1.9 本书内容

本书结构也反映出了图形系统的架构，主要按照以下方式加以组织。
第一部分内容为基础知识，如下所示。
- 图形设备。
- 几何体。
- 颜色。
- 数字图像。

第二部分内容与建模相关，如下所示。
- 场景描述。
- 几何体表达方式。
- 形式构建。
- 构成对象和层次结构。

第三部分内容则是视见机制，如下所示。
- 相机。
- 剪裁。
- 光栅化。
- 可见性。

第四部分将讨论光照，如下所示。
- 光源和材质。
- 着色（色彩化）。
- 光照模型。
- 映射。

图像合成涉及可视化和光照，因而分为两部分加以讨论。因此，计算机图形学处理过程中主要对应于四部分内容。在每一部分中，相关章节将具体处理各种组件。

前述内容曾讨论到，本书的目标是以实践方式阐述计算机图形学，因而本书将以并行方式采用如下 3 种方法。
- 计算问题分析（采用文本方式）。

- 伪代码形式的基本算法分析。
- 基于 C 语言的系统实现。

而且，每章还包含了相关练习和参考文献。

1.10 补充材料

此外，读者还可访问 http://www.visgraf.impa.br/cgtp 以了解更多信息，其中包含了本书算法实现的源代码。

第 2 章 对象和图形设备

本章将讨论图形对象的整体概念及其与图形设备间的连接方式。根据此类概念，本章将介绍用于交互式程序实现中的 2D 图形库。

2.1 图 形 对 象

对于计算机图形学的处理过程来说，理想状态下，应定义一个包容性概念，进而可对这一领域予以整体了解。相关概念基于数学模型，包括诸如几何体模型和图像等相关对象。

图像对象这一概念将是构建分析过程中的起始点，并以此将计算机图形学定义为图形对象研究领域。在计算机图形学中，相关处理包括对特定类型图形对象的操作，以及不同图形对象类型间的转换。

图形对象 $\mathcal{O}=(S,f)$[①]由子集 $S\subset\mathbb{R}^m$ 和一个函数 $f:S\to\mathbb{R}^n$ 构成。其中，S 称作 $\mathcal{O}$ 的几何支撑，并定义了图形对象的几何体和拓扑结构。另外，函数 f 指定了每个点 $p\in S$ 处的 $\mathcal{O}$ 的属性，并称作对象的属性函数，如图 2.1 所示。

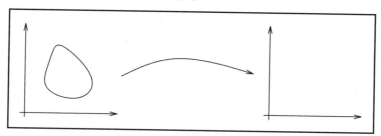

图 2.1 通用几何对象

对象 $\mathcal{O}$ 的维度由几何支撑 S 的维度指定。$\mathcal{O}$ 的多个属性对应于欧几里得 $\mathbb{R}^n$ 空间中的 1D 子空间。关于几何对象的更多信息，读者可参考[Gomes et al. 96]，其通用定义包含了针对计算机图形学中的所有相关对象，如点、曲线、表面、实体、图像和空间。

一组非常重要的图形对象由"平面对象"构成，其中 $m=2$。也就是说，几何支撑包

[①] 本书公式按照英文原版书样式编排。

含在欧几里得平面$\mathbb{R}^2$上。其原因在于，对象类可直接映射至常见的图形设备上。此类对象包含了维度$\text{Dim}(S)\leqslant 2$，并对应于点、曲线和平面区域（此处不包含分形集合）。

平面对象的两个重要例子是曲线和多边形区域。总体而言，通过适宜的方式，此类对象可用于表示曲线和平面上的任意区域。另一个重要的例子是数字图像。关于此类对象的更多信息，读者可参考[Gomes and Velho 98]。

2.1.1 图形对象的描述

从数学角度上讲，存在两种通用形式描述图形对象的几何支撑，即参数形式和隐式形式。

在参数形式中，点集$p\in S$直接由函数$g:\mathbb{R}^k\to\mathbb{R}^m$加以定义，其中，$k=\text{Dim}(S)$，如下所示。

$$S = \{(x_1,\ldots,x_m) \mid (x_1,\ldots,x_m) = g(u_1,\ldots,u_k)\}$$

在隐式描述中，S中的点采用间接方式并由函数$h:\mathbb{R}^m\to\mathbb{R}^{m-k}$确定，如下所示。

$$S = h^{-1}(c) = \{(x_1,\ldots,x_m) \mid h(x_1,\ldots,x_m) = c\}$$

【例 2.1】 （圆）当实现上述两个描述时，将使用单位圆，如图2.2所示。
- 参数描述：$(x,y)=(\sin(u),\cos(u))$，其中$u\in[0,2\pi]$。
- 隐式描述：$h^{-1}(1)$且有$h(x,y)=x^2+y^2=1$。

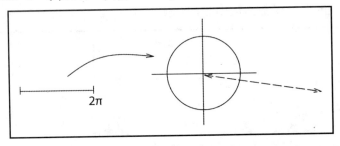

图2.2 圆的参数和隐式描述

需要注意的是，上述描述构成了图形对象几何体的连续数学模型。因此，需要获得这一类模型的有限表示，进而在计算机设备中与其协同工作，即离散机。

2.1.2 图形对象的离散化和重构

连续对象至离散对象间的转化阶段称作该对象的离散化或表示法；而相反的过程，即离散表示恢复至连续状态则称作重构。取决于整体处理过程，重构可实现准确或相似操作。

针对于此,一种较为简单的形式被广泛地应用于实践操作过程中,即基于点采样的离散化,以及基于线性插值的重构。

考查连续函数 $f: \mathbb{R} \to \mathbb{R}$。相应地,均匀点采样的表示由采样序列 $(y_i)_{i \in \mathbb{Z}}$ 加以指定。其中,$y_i = f(x_i)$ 对应于采样点 $x_i = x + i\Delta x$ 处的 f 值。通过线性插值 $\bar{f}(x) = ty_i + (1-t)y_{i+1}$,其中,$t = x \bmod \Delta x$ e $i = \lfloor x/\Delta x \rfloor$,可从采样 (y_i) 获得重构结果。注意,在当前示例中,表达结果仅提供了相似的重构,也就是说,$\bar{f} \approx f$,如图 2.3 所示。

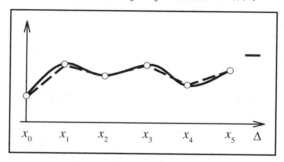

图 2.3 采样和重构

【例 2.2】 (隐式和参数对象的表达) 当构建一个圆的离散表达时(始于其参数描述),可离散化参数 $u \in [0, 2\pi]$,生成 $u_i = i/2\pi, i = 0, \ldots, N-1$,并评估 $(x_i, y_i) = g(u_i)$,进而获得圆上 N 个点的坐标。因此,圆形表达结果由这一点列表定义,如图 2.4 所示。

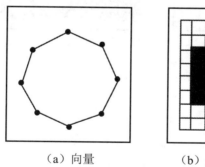 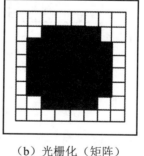

(a) 向量   (b) 光栅化(矩阵)

图 2.4 向量和光栅化(矩阵)数据格式

当构建单位圆的离散表达时(始于其隐式描述),可离散化环境空间 $\mathbb{R}^2$,并根据既定规则网格 $N \times M$ 评估隐式函数 $f(x_i, y_j)$。对应的表达结果由维度 $N \times M$ 的矩阵 A 指定。如果 $f(x_i, y_j) < 1$,则有 $a_{ij} = 1$;否则有 $a_{ij} = 0$。这一表达结果对应于圆特征函数的离散结果,如图 2.4 所示。

2.2 图形设备和表达

图形设备涵盖了一个表示空间,并于其中映射设备可操控的对象。当可视化一个图形对象 $\mathcal{O}=(U,f)$ 时,需要获得该对象的表达方式,以便这一离散对象可映射至设备的表达空间内。一旦在该空间内映射完毕,设备将执行对象的重构操作,进而生成可视化结果。

2.2.1 向量设备

在向量设备中,表达空间由点和直线段构成。更加准确地讲,表达空间表示为平面的一个子集,在这个子集中我们可以给点分配坐标。此外,给定两个点 A 和 B,设备将执行线段 AB 的重构操作。此类设备可用于可视化多边形曲线和表面,或者是多面体区域。在当前示例中,我们仅绘制多边形的边,如图 2.5(a)所示。

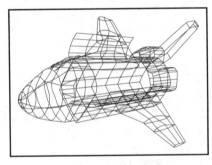

(a)向量   (b)光栅设备上

图 2.5 向量和光栅设备上的可视化

2.2.2 光栅化(矩阵)设备

此类设备的表达空间可对 $m\times n$ 矩阵进行可视化操作,其中,每个点均包含一个颜色属性,因此,当可视化这一类设备中的图形对象时,需要得到该对象的矩阵表达形式,参见[Gomes and Velho 98]。光栅化设备对于数字图像的可视化来说是较为合适的,如图 2.5(b)所示。

【例 2.3】 (渲染一个圆)对此,应采取相应的表达方式可视化(渲染)向量或光栅化设备中的图形对象。例如,圆形的可视化可通过以下方式执行:通过多边形曲线表示一个圆,如图 2.6(a)所示。当可视化光栅设备中的圆时,需要执行光栅化操作(即扫描转换）以获取其矩阵表达形式，如图 2.6（b）所示。注意，圆的多边形近似表达可显示于光栅化设备中。因此，需要对构成多边形边的直线段进行光栅化。

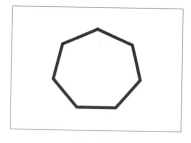 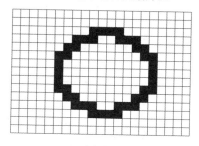

（a）向量　　　　　　　　　　　　　（b）光栅化（矩阵）

图 2.6　向量和光栅化（矩阵）表达方式

针对向量设备中的可视化操作，某些图形对象的处理过程则较为困难，甚至无法完成。多边形区域的可视化结果可通过在重构结果中放置影线标记得到。数字图像的可视化在向量设备中则较为困难。一般可采用向量格式表示结合体模型，并采用光栅化（矩阵）格式表示数字图像。

即使如此，几何体模型和数字图像可采用这两种格式中的任意一种加以表示。实际上，图形对象的概念允许对这两个元素进行统一处理。一方面，可将图像视为 Monk 表面，并在其处理过程中使用微分几何；另外一方面，可将表面的参数空间(u, v)中的坐标(x, y, z)视作图像值，通过这种方式，可针对建模操作使用图像处理技术，如图 2.7 所示。

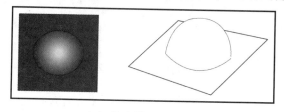 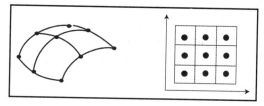

（a）作为模型的图像　　　　　　　　　　（b）作为图像的模型

图 2.7　模型和图像的统一

2.3　图形设备分类

用户-计算机与图形对象的交互可通过图形设备进行。

2.3.1 概念

当对图形设备进行分类时,可考查以下 4 种情形:物理空间、数学模型、表达方式和实现方式。因此,图形设备可根据其用途、功能、图形格式和实现结构进行分析,如图 2.8 所示。

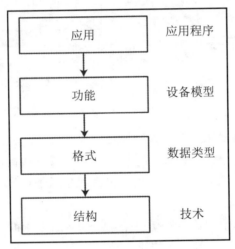

图 2.8 图形结构的抽象级别

(1)应用模式。应用模式与图形设备的应用有关。根据这一标准,图形设备可分为交互式和非交互式。

(2)功能特性。功能特性与计算模型中设备的角色相关。根据这一标准,图形设备的功能可以为输入、处理和输出。

(3)数据格式。根据表达空间的几何体特性,设备可分类为向量和光栅化(矩阵)两种。

(4)实现结构。实现结构由用于构建图形设备的拓扑结构确定,同时还涉及应用、功能和设备的格式模式。对于具有相同功能的设备,不同的实现结构示例可通过图形和矩阵显示设备予以展示。二者均为非交互式图形输出设备,但前者采用了向量格式,而后者则使用了光栅格式。

2.3.2 分类

根据上述概念,可根据具体功能对图形设备进行分类,并根据图形数据格式对每种类型进行分类。通过这种方式,可针对输入、处理和输出得到向量和光栅化(矩阵)格式的图形设备。

向量输入设备的例子包括鼠标、追踪球和游戏杆,并可通过相对坐标进行操作,而平板电脑、触摸屏和数据手套(一种可穿戴的电子设备)则采用绝对坐标进行操控。需要注意的是,除了数据手套之外,全部内容均为 2D 数据,并包含了 6 个自由度。格式化输入设备的例子包括取帧器、扫描仪和距离装置设备。

向量图形处理设备的例子包括图形管线、SGI 几何体子系统。光栅图形处理设备的例子则包括 Pixar 并行机制和 Pixel Machine。

向量图形输出设备的例子包括绘图仪和向量显示器。光栅图形输出设备的例子包括激光或喷墨打印机、CRT 或 LCD 监视器。20 世纪 60 和 70 年代,向量图形输出设备十分常见。光栅设备则在 20 世纪 80 年代占据了主导地位。当今,输入向量图形设备(例如鼠标和平板电脑)和光栅输出设备(CRT 或 LCD 监视器,以及激光打印机或喷墨打印机)往往结合在一起加以使用。

2.4 图形工作站

在实际操作过程中,图形设备往往会结合使用。对于交互式应用程序,完整的图形系统中可结合使用输入、处理和输出设备。

非交互式图形工作站是最为常见的图形系统分类。实际上,当今大多数计算设备均可视为一个图形系统。

在本书中,我们将假设图形实现是针对标准图形系统的,由光栅输出设备、矢量输入设备和通用处理器组成,如图 2.9 所示。通过图形工作站的窗口系统,向量图形输入描述将被转换为光栅描述。

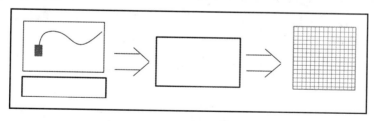

图 2.9 交互式图形工作站

2.4.1 窗口系统

现代交互式图形工作站由称之为窗口系统的图形子系统加以控制。该子系统通常与机器的操作系统进行整合，并控制图形的输入、处理和输出功能。窗口系统基于"桌面"范例，换而言之，它们实现了包含多个文档的工作表视图。在该系统类型中，每个窗口对应于独立的计算过程。窗口系统的示例包括基于 UNIX 平台的 X-Windows、基于 PC 平台的 MS-Windows，以及针对 Macintosh 平台的 Desktop。

2.4.2 视图转换

当可视化平面图形对象时，可在对象的坐标系中定义一个窗口（世界坐标系，即 WC）。该窗口应映射至设备显示空间中的一个视口中。为了提升设备的独立性，此处采用了标准化坐标系（标准化设备坐标，即 NDC）。该系统定义为[0, 1] × [0, 1]的矩形，如图 2.10 所示。

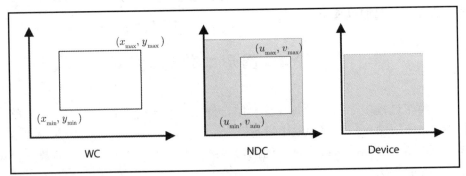

图 2.10　视图转换

在该示例中，视口定义于标准坐标系中，也正是在这个视口中映射对象空间中定义的窗口。这里，将窗口点映射至标准化坐标中的视口点这一过程称作 2D 视图转换。如果窗口通过坐标(x_{min}, y_{min})和(x_{max}, y_{max})加以定义，视口通过(u_{min}, v_{min})和(u_{max}, v_{max})加以定义，则视图转换如下所示。

$$u = \frac{u_{max} - u_{min}}{x_{max} - x_{min}}(x - x_{min}) + u_{min} \qquad (2.1)$$

$$v = \frac{v_{max} - v_{min}}{y_{max} - y_{min}}(y - y_{min}) + v_{min} \qquad (2.2)$$

在视图处理的最后阶段，视口（位于标准化坐标系中）将被映射至图形设备中。

窗口（空间坐标）、视图（标准化坐标）和图形设备间的转换是通过坐标中简单的缩放变换实现的，这将改变窗口的尺寸。

2.5 GP 图形包

与交互式图形程序相关的一个关键问题是可移植性。理想状态下,图形程序应可工作于任意平台上。至少,相同的代码应可用于各种基本类型的图形设备上。

针对这一问题的解决方案涉及设备独立性概念,其中创建了一个程序设计层,进而隔离多种设备间的实现差异。相应地,该层称作图形包,并支持不同实现的公共表达。

2.5.1 GP 特征

本书将采用 GP(图形包),该图形包最初由 Luiz Henrique Figueiredo 发布。当前 GP 版本已更新至可与 OpenGL 和 SDL 协同工作。

GP 使用了一个表达空间,且仅支持向量规范。从向量规范开始,GP 执行相应的转换操作,并重构所用设备中的图形对象。据此,可以说 GP 采用了一种向量"比喻"方式操控输入和输出中的图形对象。

总体来说,GP 假设存在一个由 2D 光栅输出设备和向量输入设备(除了键盘)组成的图形工作站。该工作站的图形系统应基于窗口范例。

下列 GP 特征体现了一种较为理想的应用状态。

- ❏ 最小化。GP 实现了一个 API(应用程序编程接口),且具有最小化特征,但对于简单的图形程序来说已经足够了。
- ❏ 可移植性。GP 是一类基于窗口范例的、与设备无关的数据包。
- ❏ 分离性。GP 体系结构隔离于图形 API 的实现细节内容。
- ❏ 可用性。该数据包支持大多数现有的平台。

当创建上述特征时,GP 的体系结构被划分为两个独立层,即 gp 层和 dv 层。

其中,gp 层负责 2D 视图转换。该级别上的例程与设备无关,其功能是将应用程序坐标映射至设备坐标中。

相应地,dv 层则负责控制图形设备。该层中的例程由 gp 层中的例程所调用。这一级别上的例程执行向量描述至设备光栅(矩阵)表达之间的转换。换而言之,该层实现了 GP 所依赖的向量"比喻"。dv 层应针对 GP 所支持的每种平台加以实现。本书并不打算对此做过多的讨论。

dv 层的当前实现针对 2D 向量图形输出采用了 OpenGL;而对窗口生成和事件管理,则使用了 SDL,其原因在于,OpenGL 和 SDL 是两种较为成熟的标准,它们实现了平台独立性，且适用于 GP 的 API 模型。OpenGL 使用了与 GP 相同的 2D 视图范例，且被大多数现代工作站中的硬件所支持。SDL（简易直接媒体层）则是一个跨平台库，并对键盘、鼠标和 3D 硬件通过 OpenGL 提供了底层支持。这在游戏领域内较为常见，且包含了与 GP 类似的事件模型。

2.5.2　GP 中的颜色属性

颜色对于任意图形对象来说都是一个十分重要的属性。在 GP 中，图形对象的属性基本上由向量图元的颜色构成。

GP 中的所采用的颜色属性处理基于颜色图（color map）这一概念，并支持 GP 表达空间中的颜色的间接定义。这里，颜色图表示为颜色空间中曲线的离散结果。其中，离散值通过一个表来表示，该表将数字索引 $i \in \{0, \dots 255\}$ 关联至设备的颜色值 $c = (r, g, b)$。其中，$r, g, b \in [0, 1]$。相应地，该表称作查找表（LUT），如图 2.11 所示。

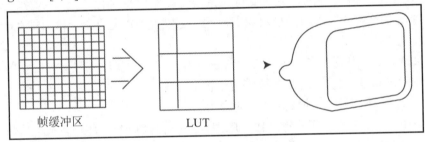

图 2.11　颜色图

该模型实现于大多数光栅显示设备中。在此类设备中，矩阵表达存储于图形板（帧缓冲区）中。每个显示单元格的颜色（像素）则通过在 LUT 中寻址而得到。

例程 gprgb 可将某个颜色与颜色图中的索引相关联。该颜色值由 R、G、B 分量的强度值所指定。如果对应属性可被成功执行，gprgb 例程将返回 1，否则返回 0。如果颜色索引包含-1 值，颜色属性将被设置为即时应用状态。为了支持全彩设备，使用 gprgb 的即时模式则较为方便，该模式在显示中实现为 24 位 RGB 值。

```
int gprgb(int c, Real r, Real g, Real b);
#define gprgb dvrgb
```

gp 中的颜色属性通过当前颜色予以执行。该颜色在 gprgb 的即时模式中被设置，或者通过 gpcolor 例程被选取。该例程的参数为一个整数，表示包含目标颜色的表输入。该例程返回上一个颜色的索引。若指定的索引无效，则返回一个负数，表示颜色图的尺寸。

```
int gpcolor(int c);
#define gpcolor dvcolor
```

2.5.3　GP 中对象的数据结构

鉴于 GP 基于窗口模型,因而 GP 中的基本图形对象表示为一个盒体。该对象由平面上的矩形构成,对应边平行于坐标轴。另外,矩形的几何体通过其主对角线的坐标(*xmin*, *ymin*),(*xmax*, *ymax*)加以定义。此外,可将一个由线性变换 $T(x, y) = (x·xu, y·yu)$ 定义的缩放属性关联到这个对象。此类缩放属性用于盒体宽高比变化,且无须重新定义整个盒体。

```
17 <Box data structure 17>≡
  typedef struct {
   Real xmin, xmax;
   Real ymin, ymax;
   Real xu, yu;
  } Box;
Defines:
 Box, used in chunk 18a.
```

因此,盒体的大小表示为 xu(xmax − xmin)和 yu(ymax − ymin)。注意,针对每个方向,xu 和 yu 表示为独立的缩放因子。

GP 的内部状态则存储于下列数据结构中。

```
18a <internal state 18a>≡
  static struct {
   Box w, v, d;
   real ax, bx, ay, by;
  } gp = {
  {0 .0, 1.0, 0.0, 1.0, 1.0, 1.0},
  {0.0, 1.0, 0.0, 1.0, 1.0, 1.0},
  {0.0, 1.0, 0.0, 1.0, 1.0, 1.0},
   1.0, 0.0, 1.0, 0.0,
  };
Defines:
 gp , used in chunks 18-21.
Uses Box 17 and real 46 46.
```

上述结构由 3 个盒体构成,即 w、v 和 d,分别表示为将可视化的场景 2D 空间的窗口、标准化坐标系中的视口和图形设备的窗口。

其中,系数 ax、bx、ay 和 by 用作缩放因子,进而实现 2D 视图转换。需要注意的是,初始状态对应于标准配置，所有盒体都是标准化的，因而视图映射为恒等函数。

GP 中的窗口包含了一个称之为背景颜色的颜色属性，然而，该属性并未存储于 Box 数据结构中。相应地，背景颜色通过查找表索引 $i = 0$ 处的颜色予以确定。如前所述，该颜色可通过例程 gppallete 或 gprgb 进行标注。

根据对应的函数，GP 中的 API 可划分为 4 个例程分类，即控制、视图、绘制和文本、图形输入和输出。稍后将详细介绍每种分类中的例程。

2.5.4 控制例程

从 GP 的初始状态可推断出，GP 仅支持单一窗口。GP 中的控制例程负责操控图形工作站屏幕上的该窗口。例程 gpopen 初始化 GP 并打开一个窗口，对应名称作为参数予以传递。

```
18b <initialization 18b>≡
  real gpopen(char* name, int width, int height)
  {
   real aspect;
   gp. d=*dvopen(name, width, height);
   calculate_aspect();
   gpwindow(0. 0,1. 0,0. 0,1. 0);
   gpviewport(0. 0,1. 0,0. 0,1. 0);
   gprgb(0,1. ,1. ,1. );
   gprgb(1,0. ,0. ,0. );
   gpcolor(1);
   return (gp. d. xu/gp. d. yu);
  }
Uses calculate aspect 19, dvopen, gp 18a, gpcolor, gprgb, and real 46 46.
```

该例程将调用 dv 层并初始化当前设备。在该调用中，设备 d 的盒体参数以及结构 gp 将被定义。除此之外，例程 gp 还将生成一个标准化窗口，并通过调用例程 gpwindow 和 gpviewport 创建一个[0, 1]×[0, 1]的视口。上述两个例程将在 2.5.5 节加以讨论。同时，例程 calculate_aspect 计算设备的盒体缩放参数，因而可映射设备中一个最大尺寸的正方形窗口。

```
19 <window aspect 19>≡
  static void calculate_aspect (void)
  {
   if (gp. d. xu > gp. d. yu) {
     gp. d. xu /= gp. d. yu;
```

```
      gp.d.yu = 1.0;
    } else {
      gp.d.yu /= gp.d.xu;
      gp.d.xu = 1.0;
    }
  }
Defines:
  calculate aspect , used in chunk 18b.
Uses gp 18a.
```

需要注意的是，例程 gpopen 将窗口的背景色初始化为白色。另外，该例程还将黑色设置为颜色图的索引 1，例程 gpcolor(1)将该颜色设置为包的当前颜色。

例程 gpclose 将关闭 GP，如果参数 wait 为正值，该过程将会等待少许时间；或者在 wait 为负值的情况下等待用户的操作。该例程实现于 dv 层中，因而定义为一个宏。

```
void gpclear(int wait);
#define gpclear dvclear
```

例程 gpclear 通过绘制背景颜色清除当前窗口。这里，参数 wait 的含义保持不变。

```
void gpclear(int wait);
#define gpclear dvclear
```

对于任意图形设备，例程 gpflush 将立即执行全部挂起操作。根据参数值 t，例程 gpwait 将暂停一段时间。若 $t > 0$，则等待 t 毫秒，否则将等待用户输入。

```
void gpflush(void);
#define gpflush dvflush

void gpwait(int t);
#define gpwait dvwait
```

2.5.5 视见例程

如前所述，例程 gpwindow 和 gpviewport 用于指定 2D 视见转换。

```
20a <window 20a>≡
  real gpwindow(real xmin, real xmax, real ymin, real ymax)
  {
    gp.w.xmin=xmin;
    gp.w.xmax=xmax;
    gp.w.ymin=ymin;
    gp.w.ymax=ymax;
```

```
    gpmake();
    dvwindow(xmin, xmax, ymin, ymax);
    return (xmax-xmin)/(ymax-ymin);
  }
Uses gp 18a, gpmake 20c, and real 46 46.

20b <viewport 20b>≡
  real gpviewport(real xmin, real xmax, real ymin, real ymax)
  {
    gp.v.xmin=xmin;
    gp.v.xmax=xmax;
    gp.v.ymin=ymin;
    gp.v.ymax=ymax;
    gpmake();
    dvviewport(xmin, xmax, ymin, ymax);
    return (xmax-xmin)/(ymax-ymin);
  }
Uses gp 18a, gpmake 20c, and real 46 46.
```

当计算窗口（场景空间）和视口（标准化坐标系）之间的转换系数时，例程 **gpwindow** 和 **gpviewport** 将调用例程 **gpmake**。

```
20c <transformation 20c>≡
  void gpmake(void)
  {
   real Ax=(gp.d.xmax-gp.d.xmin);
   real Ay=(gp.d.ymax-gp.d.ymin);
   gp.ax = (gp.v.xmax-gp.v.xmin)/(gp.w.xmax-gp.w.xmin); /* map wc to ndc */
   gp.ay = (gp.v.ymax-gp.v.ymin)/(gp.w.ymax-gp.w.ymin);
   gp.bx = gp.v.xmin-gp.ax*gp.w.xmin;
   gp.by = gp.v.ymin-gp.ay*gp.w.ymin;
   gp.ax = Ax*gp.ax; /* map ndc to dc */
   gp.ay = Ay*gp.ay;
   gp.bx = Ax*gp.bx+gp.d.xmin;
   gp.by = Ay*gp.by+gp.d.ymin;
  }
Defines:
  gpmake, used in chunk 20.
Uses gp 18a and real 46 46.
```

视见转换可通过 **gpview** 和 **gpunview** 例程高效地予以实现，这将把应用程序空间中的点映射至图形设备空间，反之亦然。

第 2 章 对象和图形设备

```
21a <view 21a>≡
  void gpview(real* x, real* y)
  {
   *x=gp.ax*(*x)+gp.bx;
   *y=gp.ay*(*y)+gp.by;
  }
Defines:
  gpview, used in chunk 24.
Uses gp 18a and real 46 46.

21b <unview 21b>≡
  void gpunview(real* x, real* y)
  {
   *x=(*x-gp.bx)/gp.ax;
   *y=(*y-gp.by)/gp.ay;
  }
Defines:
  gpunview, used in chunks 23 and 24.
Uses gp 18a and real 46 46.
```

2.5.6 绘制例程

GP 中的绘制例程负责确定显示于设备中的对象。在 GP 中，多边形曲线（开放或闭合）和多边形区域统称作多边形图元。这一类图元可通过 gpbegin、gppoint、gpend 组合予以绘制。其间，图元由调用 gppoint 给出的坐标序列定义，并通过 gpbegin 和 gpend 分隔。注意，该模式与 OpenGL 中的方式十分类似。

```
void gpbegin(int c);
#define gpbegin dvbegin

void gpend(void);
#define gpend dvend

int gppoint(Real x, Real y)
{
 gpview(&x,&y);
 return dvpoint(x,y);
}
```

具体来说，多边形图元的类型由例程 gpbegin 的参数 c 加以定义，如下所示。

- 1: 打开多边形曲线。

- p：关闭多边形曲线。
- f：填充多边形。

该模式的应用示例位于例程 gptri 的实现中，其中将绘制一个三角形区域，如下所示。

```
22 <triangle example 22>≡
  void draw_triangle(real x1, real y1, real x2, real y2, real x3, real y3)
  {
   gpbegin('f');
    gppoint(x1,y1);
    gppoint(x2,y2);
    gppoint(x3,y3);
   gpend();
  }
Defines:
  draw_triangle, never used.
Uses gpbegin, gpend, gppoint, and real 46 46.
```

文本表示为一个字母字符序列，其中，最为常见的文本属性是字符的颜色、字体类型（Helvetica、Times 等），以及字体变化（黑体、斜体等）。GP 将使用固定尺寸的字体。

下列代码表示例程 gptext 将在位置 (x, y) 处绘制字符序列 s。

```
void gptext(Real x, Real y, char* s, char* mode)
#define gptext dvtext
```

2.5.7 图形输入和交互例程

总体而言，工作站中包含了多种输入设备，其中最为常见的设备是键盘和鼠标。相应地，键盘用于字母数据的输入，而鼠标则用作一个定位器。也就是说，设备允许用户在屏幕上指定相应的位置。另外，鼠标键也可使用户定义不同的设备状态。

用户对设备的操作被系统在一个名为池化进程中捕捉。系统不断地对设备进行验证，并创建一个队列，其中每个队列输入包含设备的标识和与用户与设备交互相关的数据。相应地，该队列称作系统的事件队列。

一般情况下，gp 将鼠标和键盘用作输入设备，并通过这种方式访问事件队列，进而获得源自键盘、鼠标键和鼠标相对位置（定位器）的动作。除此之外，还存在其他一些事件可验证设备的状态（如通知窗口尺寸变化的事件）。

事件访问通过数据输入例程 gpevent 被执行，以使用户与系统进行交互。在与 GP 窗口关联的事件队列中，该例程用于检索第一个事件。其中，参数 wait 用于指定例程的行为，如下所示。

- wait! =0：等待，直至下一个事件。
- wait == 0：如果队列为空，则返回。

```
23 <event 23>≡
  char* gpevent(int wait, real* x, real* y)
  {
    int ix,iy;
    char* r=dvevent(wait,&ix,&iy);
    *x=ix; *y=iy;
    gpunview(x,y);
    return r;
  }
Defines:
 gpevent , never used.
Uses dvevent , gpunview 21b, and real 46 46.
```

上述例程根据下列代码返回事件。

- bi+：按下 i 按钮。
- bi-：释放 i 按钮。
- kt+：按下 t 键。
- ii+：按下按钮 i 时鼠标指针不移动。
- mi+：按下按钮 i 时鼠标指针移动。
- q+：窗口被窗口管理器关闭。
- r+：重绘请求。
- s+：窗口包含新尺寸(x, y)。

（1）按钮事件。在 GP 所用的标准硬件配置中，按钮设备对应于鼠标按钮。该事件的数据序列始于字符 b，随后是数字 1、2、3，表示对应的按钮。最后，+符号表示按钮被按下，-则表示按钮被释放。简而言之，按钮事件序列包含 bi+或 bi-格式。

（2）定位器事件。除了按钮设备之外，鼠标也是用于工作站中的标准定位器。鼠标移动事件始于字符 m。在当前示例中，鼠标的位置存储于例程 gpevent 的参数(x,y)中。需要注意的是，基于按钮的同步鼠标移动也始于字符 m。例如，当按钮 i 被按下时，鼠标的移动行为则表示为序列 mi+。

（3）键盘事件。当 k 键被按下时，将返回字符串 kt+，以表明当前事件。其中，t 表示对应键的 ASCII 码。

2.6 补充材料

关于平面图形对象和表示法，读者可参考[Gomes and Velho 98]以了解更多信息。另外，关于图形设备的更多信息，读者可参考[Gomes and Velho 95]。

下列例程表示 GP 的外部 API。

```
24 <API 24>≡
    real    gpopen      (char* name);
    void    gpclose     (int wait);
    void    gpclear     (int wait);
    void    gpflush     (void);
    void    gpwait      (int t);

    real    gpwindow    (real xmin, real xmax, real ymin, real ymax);
    real    gpviewport  (real xmin, real xmax, real ymin, real ymax);
    void    gpview      (real* x, real* y);
    void    gpunview    (real* x, real* y);

    int     gppalette   (int c, char* name);
    int     gprgb       (int c, real r, real g, real b);
    int     gpcolor     (int c);
    int     gpfont      (char* name);

    void    gpbegin     (int c);
    int     gppoint     (real x, real y);
    void    gpend       (void);

    void    gptext      (real x, real y, char* s, char* mode);

    char*   gpevent     (int wait, real* x, real* y);
Defines:
    gpevent, never used.
Uses gpbegin, gpclear, gpclose, gpcolor, gpend, gpflush, gppoint,
    gprgb, gptext, gpunview 21b, gpview 21a, gpwait, and real 46 46.
```

GP 图形库实现了交互式图形程序的最底层，并被限定于 2D 组件，且对应于 OpenGL 库的 2D 功能。

2.7 本章练习

（1）编译并安装 GP。

（2）利用 GP 库，编写一个交互式程序并对多边形曲线建模。其中，曲线可视为一个点列表。

（3）利用 GP 库，编写一个程序读取包含多边形曲线的文件，并对其进行绘制。

（4）涉及并实现一个工具箱界面，其中包含下列 2D 微件：按钮、评估器、选项、文本域和画布。

（5）使用练习（4）中的工具箱，编写一个程序并显示包含多个按钮的菜单。当按下按钮时，程序可在按钮上输出文本。

（6）使用练习（4）中的界面工具箱，编写一个程序并显示评估器。当评论内容被修改时，程序应可输出对应值。

（7）整合上述各项练习，针对多边形曲线实现一个完整的编辑器。其中应包含功能菜单（读取文件、写入文件、清除屏幕等）、窗口缩放和平移时的评估器，以及与鼠标键关联的编辑功能（插入一个顶点、移动一个顶点、删除一个顶点）。

（8）尝试对多边形曲线编辑器进行调整，以使其也可与 Bézier 曲线协同工作，对此，可采用细分算法对曲线的可视化结果进行细化。图 2.12 显示了当前曲线编辑器示例。

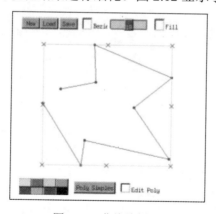

图 2.12　曲线编辑器

第 3 章 交互式图形界面

本章主要讨论交互式程序开发和用户界面设计，并构建第 2 章引入的 2D 图形包 gp 的基础设施。本章主要涉及事件处理、基于回调的界面操作、包含多个视图的交互对象、界面管理器、工具箱和微件设计。此外，还将实现一个真实的图形界面交互程序，即多边形直线编辑器。

3.1 创建交互式程序

函数 gpevent 较为简单，并可开发包含复杂和具有交互特征的图形界面。这里，交互行为由事件驱动，这可简化其实现过程。

在界面的开发过程中，需要考查两个主要元素。第一个元素是图形输入/输出。对此，应确定用户如何制定程序的各种图形对象及其行为。例如，用户可能在屏幕上简单地标记两个点，进而在绘制程序中生成一个线段。或者，用户也可能采用"橡皮筋"技术：首先定义直线段的初始点作为锚点，随后将鼠标光标拖曳至端点处（作为弹性线索）。需要注意的是，两种技术都可生成相同的结果，即生成一条直线段，但"橡皮筋"技术对于用户来说更具可控性和可见性。

界面设计也是需要考虑的另一个元素，并创建一个交互对象的全部架构，以反映程序的内部状态，以使用户可与其参数进行交互，这可通过微件和界面管理器加以实现。在前述示例的基础上，界面整体由一组按钮构成，进而可创建、删除和调整直线段。同时，它们还将与各种交互方法相关联，例如"橡皮筋"技术等。

3.2 交互基础

图形交互基本上可视为一种处理过程，在此基础上，用户通过各种逻辑命令操控对象。通常情况下，该处理过程涉及图形输出设备（如图形显示）和图形输入设备（如鼠标）间的组合。对于交互的核心内容，需要设置一个反馈循环，以便在屏幕上描述用户的动作，同时反映此类动作带来的状态变化。

3.2.1 图形反馈

图形反馈实际上整合了输入和输出,以便图形对象的行为类似于真实的实体。由用户的输入操作(如移动鼠标)引起的事件应该在图形输出方面触发相应的反应,例如,当鼠标移动时,屏幕上光标的图像也将随之发生变化。通过这种方式,用户知晓系统理解了相应的手势动作,并可查看当前的参数状态(此处为相对屏幕的鼠标位置)。

3.2.2 逻辑输入元素

第 2 章曾讨论了独立设备输出的抽象结果,以及事件的基本机制。

接下来将介绍与逻辑输入元素相关的概念,进而提供了图形输入功能。逻辑输入元素包含定位器、按钮和按键,如图 3.1 所示,这一类逻辑输入元素通常与鼠标和键盘相关联,并通过函数 gpevent 与 gp 图形包交互。

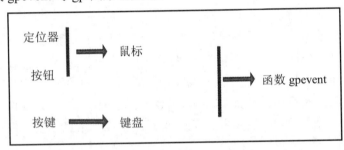

图 3.1 逻辑输入设备

定位器针对相对于窗口坐标系的 2D 坐标提供了输入内容,按钮则提供了二元状态值(即按下和释放),而按键则与 ASCII 码关联。

3.2.3 概览

接口设计的过程需要通过一些反馈实现模型将图形输入和输出整合起来,并为协调交互对象的接口管理器构建一个体系结构。

(1)反馈实现模型。最常见的反馈实现模型是池、直接事件处理、回调和装箱回调,稍后将对此类模型加以详细的讨论。

(2)界面管理器架构。界面设计架构由一个数据包构成,其中包含了针对交互式程序构建的多个元素。该数据包涵盖了以下主要组件。

- 工具箱。其中包含了一组针对各种常见任务（如从菜单中选择某个选项，或者输入文本字符串）预先包装好的微件，即界面对象。3.5 节将展示一个简单工具箱的架构和实现过程。
- 界面构造器。使用户可采用图形方式创建界面布局。
- 运行期管理器。实现持续执行过程中的反馈模型。

3.3 界面机制

为了进一步讨论和比较交互机制，本节将展示一个简单程序的伪代码，进而揭示其应用和实现过程。

3.3.1 非交互式

最简单的图形程序是非交互式的，其结构包括窗口的初始化，以及在屏幕上显示内容的一系列绘制命令。

```
main()
{
 gpopen();
 gpwindow();
 gpviewport();
 // set gpattributes
 // execute drawing primitives
 .
 .
 gpclose();
}
```

3.3.2 事件驱动

较为基础的事件驱动型交互界面使用了 gpevent 函数，明确地处理全部图形输入，并执行所关联的输出动作。

```
main()
{
 gpopen();
 gpwindow();
```

```
gpviewport();
draw_initial_state();
while (! quit) {
      e = gpevent();
      parse_exec_event(e) ;
}
gpclose();
}
```

需要注意的是，主要的实现内容位于 parse_exec_event 函数中，该函数负责明确地处理所有交互行为。

```
parse_exec_event(e)
{
 switch (e) {
 case k: // key pressed
     .
     .
 case m: // mouse movement
     .
     .
     .
 }
}
```

鉴于程序越来越复杂，其间涉及更多的接口，因而该模型将变得难于扩展和维护。其原因在于，每个输入事件须明确地被处理，同时兼顾考查受到影响的对象和程序状态。例如，当按下某个键时，它可能具有不同的含义，具体取决于鼠标所在的位置和所选择的对象。

3.3.3 回调模型

回调模型解决了直接处理输入事件所带来的困难。该模型使用事件，但是须将特定的事件关联到特定的图形对象或接口条件中。

例如，特定的函数可与按下鼠标左键这一动作相关联。在该模型中，当鼠标左键被按下时将调用该函数。因此，该函数被称作回调函数。

```
main()
{
 gpopen();
 gpwindow();
 gpregister("b1+", f1, d1);
```

```
    gpregister("b2+", f2, d2);
      .
      .
    gpmain_loop ()
    gpclose()
}
```

因此，在初始化阶段，用户通过 gpregister 函数定义全部回调动作。随后，交互循环通过函数 gpmain 循环加以实现，并通过在适当的时间调用所需的动作来自动处理事件。通过这种方式，交互行为可以通过简单地替换回调的实现来更改。

3.3.4 包含多个视图的回调

通过构建事件和图形对象间的链接，可极大地改进回调模型。需要注意的是，在通用回调模型中，事件关联具有全局特征。也就是说，同一回调可针对特定的事件类被激活，如按下鼠标键。

包含多个视图的回调可将本地事件关联至某个动作上，例如，取决于哪一个鼠标键被按下，不同的回调将被激活。

该模型在多个视图的帮助下得以实现。屏幕被不同的区域平铺而成，同时包含了不同的本地动作。

考查下列函数：

```
mvreg(1,"b1+",displ1,id1);
```

其中，如果鼠标键 1 在屏幕区域 v1 中被按下，则回调函数 displ1(id1)将被激活。下列函数则在屏幕区域 v2 中指定了类似的动作 displ2：

```
mvreg(2,"b1+",displ2,id2);
```

当然，在该模型中，也可维护全局事件。这可针对整个屏幕并通过一个特定的标识符（-1）加以实现，如下所示。

```
mvreg(-1,"=q",exit,0); // call exit(0) if key 'k' is pressed in any area
```

包含多个视图的回调是我们将要采用的模型，进而构建工具箱基础设施。在该模型的基础上，交互式程序的结构如下所示。

```
main()
{
 gpopen()
 mvopen()
```

```
interface_setup()
mvmain_loop ()
gpclose(0)
}
```

界面的配置通过 interface_setup 函数完成,这定义了每个视图区域、对应的回调以及界面的初始状态。

```
interface_setup()
{
  mvviewport(1, x, x, y, y)
  .
  mvregister (1, , x, 0)
  .
  draw_initial_state()
}
```

3.4 节将描述多视图回调模型的实现。

3.4 界面对象

图形界面对象可利用 3.3 节讨论的多视图回调模型予以创建。多视口框架使得界面对象可与屏幕区域进行关联,通过事件驱动型图形反馈,回调框架使得此类对象处于激活状态。

mvcb 包提供了此类框架的集成实现。

3.4.1 多视口

实际上,多视口框架在 gp 上提供了一个平铺的屏幕管理器,且需要实现多视图的抽象内容。每个视图的行为类似于 gp 包,但仅局限于某个特定的屏幕区域。

mv 的内部状态由一组视图构成,每个视图通过窗口和视口加以定义。除此之外,还存在一个"当前视图"的概念,所有的 gp 命令都适用于这一概念。

```
32 <mv internalstate 32>≡
  static int    nv;       /* number of views */
  static Box*   w;        /* windows */
  static Box*   v;        /* viewports */
  static int    current;  /* current view */
Defines:
```

```
current, used in chunks 33-35.
nv, used in chunks 33, 35b, and 36a.
v, used in chunks 33, 35-39, 41c, and 44.
w, used in chunks 33-35 and 42-44.
```

gp 的主要控制功能在 mv 包中被复制，并封装了对应的功能。

33a `<mv open 33a>≡`
```
  int mvopen(int n)
  {
  if (n<=0) return 0;
  v=(Box*) emalloc(n*sizeof(Box)); if (v==0) return 0;
  w=(Box*) emalloc(n*sizeof(Box)); if (w==0) return 0;
  nv=n;
  current=0;
  for (n=0; n<nv; n++) {
   w[n].xu = w[n].yu = 1.0;
   mvwindow(n,0.0,1.0,0.0,1.0);
   mvviewport(n,0.0,1.0,0.0,1.0);
  }
  return 1;
}
Defines:
  mvopen, used in chunk 40b.
Uses current 32, mvviewport 33d, mvwindow 33c, nv 32, v 32, and w 32.
```

33b `<mv close 33b>≡`
```
  void mvclose(void)
  {
   efree(w);
   efree(v);
  }
Defines:
  mvclose, used in chunk 41a.
Uses v 32 and w 32.
```

33c `<mv window 33c>≡`
```
  void mvwindow(int n, real xmin, real xmax, real ymin, real ymax)
  {
   if (n<0| | n>=nv) return;
   w[n].xmin=xmin;
   w[n].xmax=xmax;
   w[n].ymin=ymin;
```

```
      w[n].ymax=ymax;
    }
Defines:
  mvwindow, used in chunks 33a and 42a.
Uses nv 32, real 46 46, and w 32.

33d <mv viewport 33d>≡
  void mvviewport(int n, real xmin, real xmax, real ymin, real ymax)
    {
    if (n<0| | n>=nv) return;
    v[n].xmin=xmin;
    v[n].xmax=xmax;
    v[n].ymin=ymin;
    v[n].ymax=ymax;
    }
Defines:
  mvviewport, used in chunks 33a, 34c, and 42a.
Uses nv 32, real 46 46, and v 32.

34a <mv clear 34a>≡
  void mvclear(int c)
    {
    int old=gpcolor(c);
    int n=current;
    gpbox(w[n].xmin,w[n].xmax,w[n].ymin,w[n].ymax);
    gpcolor(old);
    }
Defines:
  mvclear, never used.
Uses current 32 and w 32.
```

辅助函数 mvframe 围绕视图绘制一条轮廓线,进而可方便地在屏幕上看到视图的区域。

```
34b <mv frame 34b>≡
  void mvframe(void)
    {
    int n = current;
    gpline(w[n].xmin,w[n].ymin,w[n].xmax,w[n].ymin);
    gpline(w[n].xmax,w[n].ymin,w[n].xmax,w[n].ymax);
    gpline(w[n].xmax,w[n].ymax,w[n].xmin,w[n].ymax);
    gpline(w[n].xmin,w[n].ymax,w[n].xmin,w[n].ymin);
    }
Defines:
```

第 3 章 交互式图形界面

```
  mvframe, used in chunk 43b.
Uses current 32 and w 32.
```

函数 mvdiv 将屏幕的矩形区域划分为 nx×ny 个视图单元。

34c <mv divide 34c>≡
```
  void mvdiv(int nx, int ny, real xvmin, real xvmax, real yvmin, real yvmax)
  {
   int i,n;
   real dx=(xvmax-xvmin)/nx;
   real dy=(yvmax-yvmin)/ny;
   for (n=0,i=0; i<ny; i++)
   {
    int j;
    real ymax=yvmax-i*dy;
    real ymin=ymax-dy;
    for (j=0; j<nx; j++,n++) {
     real xmin=xvmin+j*dx;
     real xmax=xmin+dx;
     mvviewport(n,xmin,xmax,ymin,ymax);
    }
   }
  }
Defines:
  mvdiv, used in chunk 35a.
Uses mvviewport 33d and real 46 46.
```

函数 mvmake 针对整个屏幕区域使用了 mvdiv。

35a <mv make 35a>≡
```
  void mvmake(int nx, int ny)
  {
   real x,y;
   if (nx>ny) {
    x=1.0;
    y=((real)ny)/nx;
   } else {
    x=((real)nx)/ny;
    y=1.0;
   }
   mvdiv(nx,ny,0.0,x,0.0,y);
   gpviewport(0.0,x,0.0,y);
  }
Defines:
```

```
    mvmake, never used.
Uses mvdiv 34c and real 46 46.
```

函数 mvact 使得指定的视图处于激活状态，也就是说，变为当前视图。

```
35b <mv activate 35b>≡
  int mvact(int n)
  {
  int old=current;
  if (n<0| | n>=nv) return old;
  gpwindow(w[n].xmin,w[n].xmax,w[n].ymin,w[n].ymax);
  gpviewport(v[n].xmin,v[n].xmax,v[n].ymin,v[n].ymax);
  current=n;
  return old;
  }
Defines:
  mvact, used in chunks 36a and 43b.
Uses current 32, nv 32, v 32, and w 32.
```

3.4.2　基于视图的回调

回调模型主要针对多个视图加以实现，即创建某种机制，并将事件与视图关联。对此，可定义一个名为 mvevent 的函数。

```
36a <mv event 36a>≡
  char* mvevent(int wait, real* x, real* y, int* view)
  {
  int n; real gx,gy, tx,ty;
  char* r=gpevent(wait,&gx,&gy);
  if (r==NULL) return r;
  gpview(&gx,&gy); tx=gx; ty=gy;
  gpwindow(0.0,1.0,0.0,1.0);
  gpviewport(0.0,1.0,0.0,1.0);
  gpunview(&gx,&gy);
  *view=-1;
  for (n=0; n<nv; n++) {
    if (gx>=v[n].xmin && gx<=v[n].xmax && gy>=v[n].ymin && gy<=v[n].ymax) {
      int old=mvact(n);
      gpunview(&tx,&ty);
      *x=tx;
      *y=ty;
      *view=n;
```

```
      mvact(old);
      break;
    }
  }
  return r;
}
```
Defines:
  mvevent, used in chunk 38a.
Uses mvact 35b, nv 32, real 46 46, and v 32.

回调抽象通过与视图匹配的事件模式列表加以实现。

36b <mv callbacks state 36b>≡
```
  typedef struct event Event;

  struct event {
    int v;
    char* s;
    MvCallback* f;
    void* d;
    Event* next;
  };

  static Event* firstevent=NULL;
  static int gp_wait=1;
```
Defines:
  firstevent, used in chunk 37.
  gp_wait, used in chunks 37b and 38a.
Uses next 37a 46 and v 32.

出于方便，这里定义了下列宏：

37a <mvcb macros 37a> ≡
```
  #define new(t)             ( (t*) emalloc(sizeof(t)) )
  #define streq(x,y)         (strcmp(x,y)==0)
  #define V(_)               ((_)->v)
  #define S(_)               ((_)->s)
  #define F(_)               ((_)->f)
  #define D(_)               ((_)->d)
  #define next(_)            ((_)->next)
  #define foreachevent(e) for (e=firstevent; e!=NULL; e=next(e))

  static Event*     findevent    (int v, char* s);
```

```
  static Event*     matchevent   (int v, char* s);
  static int        match        (char *s, char *pat);
Defines:
  D, used in chunks 37b and 38a.
  F, used in chunks 37b and 38a.
  findevent, used in chunk 37b.
  foreachevent, used in chunks 38b and 39a.
  matchevent, used in chunk 38a.
  new, used in chunks 37b and 47.
  next, used in chunks 36b, 37b, and 47.
  S, used in chunks 37-39.
  streq, used in chunk 38b.
  V, used in chunks 37-39.
Uses firstevent 36b, match 39b, and v 32.
```

函数 mvregister 将回调动作关联至特定的事件和视图上。

```
37b <mv register function 37b>≡
  MvCallback* mvregister(int v, char* s, MvCallback* f, void* d)
  {
   MvCallback* old;
   Event* e=findevent(v,s);
   if (e==NULL) {
    static Event* lastevent=NULL;
    e=new(Event);                          /* watch out for NULL! */
    V(e)=v;
    S(e)=s;
    F(e)=NULL;
    next(e)=NULL;
    if (firstevent==NULL) firstevent=e; else next(lastevent)=e;
    lastevent=e;
   }
   old=F(e);
   F(e)=f;
   D(e)=d;
   if (s[0] =='i' && f! =NULL) gp_wait=0;
   return old;
  }
Defines:
  mvregister, used in chunks 40b and 43a.
Uses D 37a, F 37a, findevent 37a 38b, firstevent 36b, gp wait 36b, new 37a
46, next 37a 46, S 37a, V 37a, and v 32.
```

函数 mvmainloop 实现了事件与视图匹配的运行期回调模型。

```
38a <mv mainloop 38a>≡
  void mvmainloop(void)
  {
    for (; ; ) {
      real x,y;
      int v;
      char* s=mvevent(gp_wait,&x,&y,&v);
      Event*e=matchevent(v,s);
      if (e!=NULL && F(e)(D(e),v,x,y,s))
        break;
    }
  }
Defines:
  mvmainloop, used in chunk 41b.
Uses D 37a, F 37a, gp_wait 36b, matchevent 37a 39a, mvevent 36a, real 46
46, and v 32.
```

当处理某个事件时，函数 findevent 和 matchevent 用于查询事件模式列表。

```
38b <find event 38b>≡
  static Event* findevent(int v, char* s)
  {
    Event* e;
    foreachevent(e) {
      if (V(e)!=v) continue;
      if (s==NULL && S(e)==NULL) break;
      if (s==NULL|| S(e)==NULL) continue;
      if (streq(S(e),s)) break;
    }
    return e;
  }
Defines:
  findevent, used in chunk 37b.
Uses foreachevent 37a, S 37a, streq 37a, V 37a, and v 32.

39a <match event 39a>≡
  static Event* matchevent(int v, char* s)
  {
    Event* e;
    foreachevent(e) {
      if (V(e)<0|| V(e)==v)
```

```
        if (match(S(e),s)) break;
    }
    return e;
}
Defines:
  matchevent, used in chunk 38a.
Uses foreachevent 37a, match 39b, S 37a, V 37a, and v 32.
```

实际的字符串模式匹配通过辅助函数 match 完成。

```
39b <match string 39b> ≡
 static int match(char *s, char *pat)
 {
  if (s==NULL) return pat==NULL;
  if (pat==NULL) return s==NULL;
  for (; *s! =0; s++, pat++) {
   if (*s! =*pat) return 0;
  }
  return 1;
 }
Defines:
  match, used in chunks 37a and 39a.
```

3.5　工　具　箱

tk 工具包构建于 mvcb 工具包之上，并创建界面对象（即微件）。屏幕的矩形区域将与此类对象关联，tk 库针对每种微件类型实现了相应的反馈。这是通过针对每个活动微件注册特定的回调而实现的。

例如，pushbutton 微件包含一个二元状态值（开/关），并在屏幕上表示为一个框体，并在黑色或白色背景（取决于当前值）上设置了文本内容。每次用户单击该按钮时，其状态将会发生变化；同时，回调函数通知用户程序对应值已发生变化。需要注意的是，图形反馈通过微件被自动处理。

一般来说，工具箱创建了一个抽象层，并实现了应用程序所用的基本界面对象。

3.5.1　基本元素

界面工具箱设计中的核心问题是所实现的微件集合的定义，以及新微件的创建机制。此处建议使用一个最小的微件集，以实现通用用户界面的必要功能。具体来说，最小工具箱由以下微件构成：按钮、滑块、选项、文本域和图形画布。图 3.2 显示了每种微件的简单的图形描述。

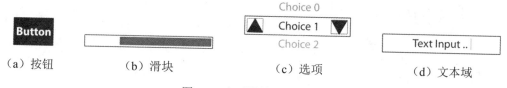

（a）按钮　　　　（b）滑块　　　　（c）选项　　　　（d）文本域

图 3.2　必要微件的图形描术

3.5.2　tk 包

微件 API 由创建和销毁微件实例的函数构成，并将其映射/反映射至屏幕上。对应函数如下所示。

```
w = create_widget (pos, par, fun)
destroy_widget (w)
map_widget (w)
unmap_widget (w)
```

包的内部状态包含一个微件指针向量、向量的大小以及向量中的最后一个有效项。

```
40a <tk local state 40a>≡
  Widget **wa = NULL;
  int wn = 0;
  int wi = 0;
Defines:
  wa, used in chunks 40-42.
  wi, used in chunks 40-42.
  wn, used in chunks 40-42.
Uses Widget 44c.
```

运行期界面管理器的基本功能通过 tk_open 函数予以实现，该函数将初始化界面；函数 tk_close 将终止界面；tk_mainloop 函数则用于处理交互循环。

```
40b <tk initialization 40b>≡
  void tk_open(int n)
  {
    int i;
    mvopen(n);
    wa = NEWARRAY(n, Widget *);
    for (i=0; i<n; i++)
      wa[i] = NULL;
```

```
    wn = n;
    wi = 0;
    mvregister(-1,"r",tk_redraw,NULL);
    gpflush();
  }
Defines:
  tk open, never used.
Uses mvopen 33a, mvregister 37b, tk redraw 41c, wa 40a, wi 40a, Widget 44c,
and wn 40a.
```

41a <tk close 41a>≡
```
  void tk_close()
  {
    efree(wa);
    wa = NULL; wn = wi = 0;
    mvclose();
  }
Defines:
  tk close, never used.
Uses mvclose 33b, wa 40a, wi 40a, and wn 40a.
```

41b <tk main loop 41b>≡
```
  void tk_mainloop()
  {
    mvmainloop();
  }
Defines:
  tk mainloop, never used.
Uses mvmainloop 38a.
```

函数 tk_redraw 用于在屏幕上显示当前处于活动状态下的全部微件。

41c <tk redraw 41c>≡
```
  int tk_redraw(void* p, int v, real x, real y, char* e)
  {
    int i;
    fprintf(stderr, "redraw\n"); fflush(stderr);
    for (i=0; i<wi; i++) {
      switch (wa[i] ->type) {
      case TK_BUTTON:
        button_draw(wa[i], 1); break;
      default:
        error("tk"); break;
```

第 3 章 交互式图形界面

```
    }
  }
  gpflush();
  return 0;
}
```
Defines:
  tk redraw, used in chunk 40b.
Uses button draw 43b, real 46 46, redraw 47, TK BUTTON, v 32, wa 40a, and wi 40a.

新的微件通过调用 tk_widget 函数指定其类型和参数并进行实例化。

42a `<tk widget 42a>≡`
```
  Widget* tk_widget(int type, real x, real y, int (*f)(), void *d)
  {
    Widget *w = widget_new(type, x, y, 0.2, f);
    if (wi >= wn)
      error("tk");
    w->id = wi++;
    wa[w->id] = w;
    mvwindow(w->id, 0, 1, 0, 1);
    mvviewport(w->id, w->xo, w->xo + w->xs, w->yo, w->yo + w->ys);
    switch (type) {
    case TK_BUTTON:
      button_make(w, d); break;
    default:
      error("tk"); break;
    }
    return w;
  }
```
Defines:
  tk_widget, never used.
Uses button_make 43a, mvviewport 33d, mvwindow 33c, real 46 46, TK_BUTTON, w 32, wa 40a, wi 40a, Widget 44c, widget new 42b, and wn 40a.

内部函数 widget_new 创建一个通用微件对象，并于随后绑定至特定的微件类上。

42b `<new widget 42b>≡`
```
  Widget* widget_new(int type, real x, real y, real s, int (*f)())
  {
    Widget *w = NEWSTRUCT(Widget);
    w->id = -1;
    w->type = type;
    w->xo = x; w->yo = y;
```

```
    w->xs = w->ys = s;
    w->f = f;
    w->d = NULL;
    return w;
  }
Defines:
  widget_new, used in chunk 42a.
Uses real 46 46, w 32, and Widget 44c.
```

新的微件类在 tk 框架中被定义,即针对创建、绘制以及交互机制指定相应的函数,并通过 mvcb 包中的回调进行处理。

作为新文件类的构建示例,下面将展示如何定义一个按钮微件,这可通过 button_make 和 button_draw 加以实现。

```
43a <make button 43a>≡
  void button_make(Widget *w, char *s)
  {
    mvregister(w->id,"b1+",button_pressed,w);
    mvregister(w->id,"b1-",button_released,w);
    w->d = s;
    button_draw(w, 1);
  }
Defines:
  button make, used in chunk 42a.
Uses button draw 43b, button pressed 44a, button_released 44b, mvregister
37b, w 32, and Widget 44c.

43b <draw button 43b>≡
  void button_draw(Widget *w, int posneg)
  {
    char *label = w->d;
    int fg, bg;
    if (posneg) {
      fg = 1; bg = 0;
    } else {
      fg = 0; bg = 1;
    }
    mvact(w->id);
    gpcolor(fg);
    gpbox(0., 1., 0., 1.);
    gpcolor(bg);
    gptext(.2, .2, label, NULL);
```

第 3 章 交互式图形界面

```
    mvframe();
    gpflush();
  }
Defines:
  button_draw, used in chunks 41c, 43, and 44.
Uses mvact 35b, mvframe 34b, w 32, and Widget 44c.
```

按钮的行为通过回调函数 button_pressed 和 button_released 加以控制,它们将分别处理按钮的按下和释放操作。

```
44a <press action 44a>≡
  int button_pressed(void* p, int v, real x, real y, char* e)
  {
    button_draw(p, 0);
    return 0;
  }
Defines:
  button_pressed, used in chunk 43a.
Uses button_draw 43b, real 46 46, and v 32.

44b <release action 44b>≡
  int button_released(void* p, int v, real x, real y, char* e)
  {
    Widget *w = p;
    button_draw(w, 1);
    return w->f();
  }
Defines:
  button_released, used in chunk 43a.
Uses button_draw 43b, real 46 46, v 32, w 32, and Widget 44c.
```

微件对象通过包含其 ID、类型、位置、屏幕尺寸、本地数据和应用程序回调函数的数据结构加以定义。

```
44c <widget data structure 44c>≡
  typedef struct Widget {
    int id;
    int type;
    real xo, yo;
    real xs, ys;
    void* d;
    int (*f)();
  } Widget;
```

```
Defines:
  Widget, used in chunks 40 and 42-44.
Uses real 46 46.
```

3.5.3 示例

作为一个使用 tk 工具箱的图形交互式程序生成其界面的示例,下面将显示一个简单的应用程序,并在屏幕上创建两个按钮。其中一个按钮负责输出值;另一个按钮则用于退出程序。图 3.3 显示了该程序的界面布局。

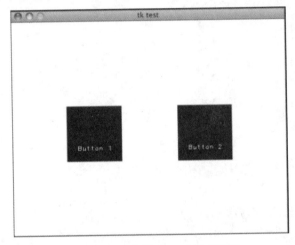

图 3.3 使用 tk 的交互式程序示例

```
int main(int argc, char* argv[] )
{
  Widget *w0;
  gpopen("tk test", 512, 512);
  tk_open(10);
  tk_widget(TK_BUTTON, .2, .5, but1, "Button 1");
  tk_widget(TK_BUTTON, .6, .5, but2, "Button 2");
  tk_mainloop();
  tk_close();
  gpclose(0);
}

int but1()
{
```

```
    fprintf(stderr, "Button 1 pressed\n"); fflush(stderr);
    return 0;
}

int but2()
{
    fprintf(stderr, "Button 2 pressed - quitting\n"); fflush(stderr);
    return 1; // exits the main loop when 1 is returned.
}
```

3.6 多边形直线编辑器

作为一个图形画布（canvas）的应用示例，本节将展示多边形直线编辑器的实现过程。需要注意的时候，该程序针对直线输入实现了一种"橡皮筋"方法，前述内容曾对此有所介绍。图 3.4 显示了该应用程序的屏幕内容。

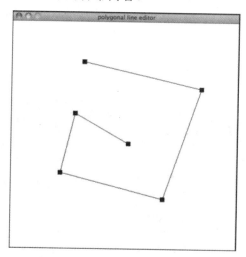

图 3.4　多边形直线编辑器

```
46 <ple state 46>≡

  #define TOL tol

  typedef struct point Point;
```

```
struct point {
 real x,y;
 Point* next;
 Point* prev;
};

void    redraw          (int clear);
void    delpoints       (void);
void    showpolygon     (void);
void    showspline      (void);
void    showpoints      (void);
void    addpoint        (real x, real y);
void    movepoint       (real x, real y);
void    delpoint        (real x, real y);
void    startmove       (real x, real y);
void    endmove         (real x, real y);
void    showchange      (Point* p, int c);
void    showpoint       (Point* p);
void    showside        (Point* p, Point *q);
Point*  findpoint       (real x, real y);

Callback
 do_polygon,
 do_quit,
 do_redraw,
 do_addpoint,
 do_startmove,
 do_endmove,
 do_delpoint,
 do_movepoint;

#define X(p)        ((p)->x)
#define Y(p)        ((p)->y)
#define new(t)      ((t*)emalloc(sizeof(t)))
#define next(p)     ((p)->next)
#define prev(p)     ((p)->prev)

static Point*       firstpoint=NULL;
static Point*       lastpoint=NULL;
static Point*         moving=NULL;
static int              showingpolygon=1;
static int              showingpoints=1;
```

```
    static real              xmin = 0, xmax = 1, ymin = 0, ymax = 1;
    static real              aspect = 1, tol = 0.1;
Defines:
    findpoint, never used.
    firstpoint, used in chunk 47.
    lastpoint, used in chunk 47.
    moving, used in chunk 47.
    new, used in chunks 37b and 47.
    next, used in chunks 36b, 37b, and 47.
    prev, used in chunk 47.
    real, used in chunks 18, 20-24, 33-36, 38a, 41, 42, 44, and 47.
    showingpoints, used in chunk 47.
    showingpolygon, used in chunk 47.
    TOL, used in chunk 47.
    X, used in chunk 47.
    Y, used in chunk 47.
Uses addpoint 47, delpoint 47, delpoints 47, do_addpoint 47, do_delpoint
47, do_endmove 47, do_movepoint 47, do_polygon 47, do_quit 47, do_redraw
47, do_startmove 47, endmove 47, movepoint 47, redraw 47, showchange 47,
showpoint 47, showpoints 47, showpolygon 47, showside 47, and startmove 47.

47 <ple functions 47>≡
  int main(int argc, char* argv[] )
  {
    gpopen("polygonal line editor", 512 * aspect, 512);
    gpwindow(xmin,xmax, ymin,ymax);

    gpmark(0,"B"); /* filled box mark */
    gpregister("kp",do_polygon,0);
    gpregister("kq",do_quit,0);
    gpregister("kr",do_redraw,0);
    gpregister("k\f",do_redraw,0);
    gpregister("b1+",do_addpoint,0);
    gpregister("kd",do_delpoint,0);
    gpregister("b3+",do_startmove,0);
    gpregister("b3-",do_endmove,0);
    gpregister("m3+",do_movepoint,0);

    gpmainloop();
    gpclose(0);
  }
```

```
void redraw(int clear)
{
 if (clear)
   gpclear(0);
 if (showingpolygon)
   showpolygon();
 showpoints();
 gpflush();
}

void delpoints(void)
{
 firstpoint=lastpoint=NULL; /* lazy! */
}

void addpoint(real x, real y)
{
 Point* p=new(Point);
 X(p)=x;
 Y(p)=y;
 next(p)=NULL;
 if (showingpoints) showpoint(p);
 if (firstpoint==NULL) {
  prev(p)=NULL;
  firstpoint=p;
 } else {
  prev(p)=lastpoint; next(lastpoint)=p;
  if (showingpolygon) showside(lastpoint,p);
 }
 lastpoint=p;
}

void delpoint(real x, real y)
{
 Point* p=findpoint(x,y);
 if (p!=NULL) {
  if (prev(p)==NULL) firstpoint=next(p); else next(prev(p))=next(p);
  if (next(p)==NULL) lastpoint=prev(p); else prev(next(p))=prev(p);
  redraw(1);
 }
}
```

第3章 交互式图形界面

```
void startmove(real x, real y)
{
 moving=findpoint(x,y);
 if (moving! =NULL) {
  x=X(moving); y=Y(moving);
  gpcolor(0); gpplot(x,y); gpcolor(1);
  gpmark(0,"b"); gpplot(x,y);
 }
}

void movepoint(real x, real y)
{
 if (moving! =NULL) {
  showchange(moving,0);
  X(moving)=x; Y(moving)=y;
  showchange(moving,1);
 }
 else startmove(x,y);
}

void endmove(real x, real y)
{
 if (moving! =NULL) {
  gpmark(0,"B");
  redraw(0);
  moving=NULL;
 }
}

Point* findpoint(real x, real y)
{
 Point* p=firstpoint;
 for (p=firstpoint; p! =NULL; p=next(p)) {
  if ((fabs(X(p)-x)+fabs(Y(p)-y))<TOL) break;
 }
 return p;
}

void showpoints(void)
{
 Point* p;
 for (p=firstpoint; p! =NULL; p=next(p))
```

```
   showpoint(p);
 gpflush();
}

void showpolygon(void)
{
 Point* p;
 for (p=firstpoint; p!=NULL; p=next(p))
   showside(p,next(p));
 gpflush();
}

void showpoint(Point* p)
{
 gpplot(X(p),Y(p));
}

void showside(Point* p, Point *q)
{
 if (p!=NULL && q!=NULL) gpline(X(p),Y(p),X(q),Y(q));
}

void showchange(Point* p, int c)
{
 gpcolor(c);
 showpoint(p);
 if (showingpolygon) {
   showside(prev(p),p);
   showside(p,next(p));
 }
 gpflush();
}

int do_clear(char* e, real x, real y, void* p)
{
 delpoints();
 redraw(1);
 return 0;
}

int do_polygon(char* e, real x, real y, void* p)
 {
```

```
 showingpolygon=!showingpolygon;
 redraw(1);
 return 0;
}

int do_quit(char* e, real x, real y, void* p)
{
 return 1;
}

int do_redraw(char* e, real x, real y, void* p)
{
 redraw(1);
 return 0;
}

int do_addpoint(char* e, real x, real y, void* p)
{
 addpoint(x,y);
 gpflush();
 return 0;
}

int do_startmove(char* e, real x, real y, void* p)
{
 startmove(x,y);
 gpflush();
 return 0;
}

int do_endmove(char* e, real x, real y, void* p)
{
 endmove(x,y);
 gpflush();
 return 0;
}

int do_delpoint(char* e, real x, real y, void* p)
{
 delpoint(x,y);
 gpflush();
 return 0;
```

```
    }

    int do_movepoint(char* e, real x, real y, void* p)
    {
     movepoint(x,y);
     gpflush();
     return 0;
    }

Defines:
  addpoint, used in chunk 46.
  delpoint, used in chunk 46.
  delpoints, used in chunk 46.
  do_addpoint, used in chunk 46.
  do_clear, never used.
  do_delpoint, used in chunk 46.
  do_endmove, used in chunk 46.
  do_movepoint, used in chunk 46.
  do_polygon, used in chunk 46.
  do_quit, used in chunk 46.
  do_redraw, used in chunk 46.
  do_startmove, used in chunk 46.
  endmove, used in chunk 46.
  findpoint, never used.
  main, used in chunks 313, 317, and 318c.
  movepoint, used in chunk 46.
  redraw, used in chunks 41c and 46.
  showchange, used in chunk 46.
  showpoint, used in chunk 46.
  showpoints, used in chunk 46.
  showpolygon, used in chunk 46.
  showside, used in chunk 46.
  startmove, used in chunk 46.
Uses firstpoint 46, lastpoint 46, moving 46, new 37a 46, next 37a 46, prev
46, real 46 46, showingpoints 46, showingpolygon 46, TOL 46, X 46, and Y 46.
```

3.7 回　　顾

本章讨论了界面设计的架构，其中包含了 4 层内容，如图 3.5 所示。第一层为图形界面程序；第二层为界面工具箱，并由 tk 包予以实现；第三层为 gp 包实现的图形输入和输出；第四层为窗口系统且与平台无关，如 Linux 平台中的 X11，Microsoft Windows 平台中的 Vista，以及 MacOS X 平台中的 Aqua/Cocoa。

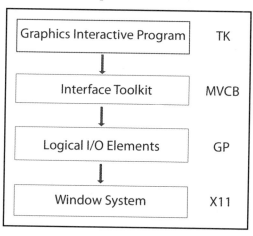

图 3.5　实现层

3.8　补 充 材 料

本章针对计算机图形学中的界面设计讨论了库的实现。除此之外，其他一些较为流行的工具箱库还包括 QT、GTK、FLTK 和 GLUI。

MVCB 库的外部 API 由下列代码构成：

```
int        mvopen    (int n);
void       mvclose   (void);
void       mvwindow  (int n, real xmin, real xmax, real ymin, real ymax);
void       mvviewport(int n, real xmin, real xmax, real ymin, real ymax);
int        mvact     (int n);
void       mvclear   (int c);
void       mvframe   (void);
void       mvmake    (int nx, int ny);
void       mvdiv     (int nx, int ny, real xmin, real xmax, real ymin, real ymax);

char*      mvevent   (int wait, real* x, real* y, int* view);

void       mvmainloop(void);
MvCallback* mvregister(int v, char* s, MvCallback* f, void* d);
```

3.9 本章练习

（1）在 tk 库中整合映射和反映射操作。
（2）扩展 tk 库，以使其包含一个滑块微件。
（3）扩展 tk 库，以使其包含一个选择微件。
（4）扩展 tk 库，以使其包含一个文本微件。

第 4 章 几 何 体

本章将介绍计算机图形学中的几何体,并尝试开发一个计算工具,进而提供多种图形学问题的解决方案。

4.1 计算机图形学中的几何体

第一个问题是,在计算机图形学中,什么是最为适宜的几何体?为了回答这一问题,我们需要考查需要解决的问题类型以及计算方面的内容。

4.1.1 应用和功能

几何体以多种方式呈现于各种计算机图形学处理过程中。

在建模过程中,几何体在模型操控、属性计算方面用于呈现建模对象的形式。3D 对象建模的自然空间表示为 $\mathbb{R}^3$。

在视图机制中,几何体用于虚拟相机、光照模拟和图像生成的描述过程。这一类问题涉及光学、投影和坐标系间的转换操作。通过这种方式,视图处理包含 3D 场景空间和图像的 2D 空间。

此外,动画设计还将包括一段时间内多个场景参数的转换操作。

4.1.2 计算内容

针对计算过程来说,几何元素和几何操作涵盖了不同的应用。相应地,几何元素应包含简单和自然的描述,部分原因在于:应可从基本的元素构建复杂的元素。几何操作应针对几何元素的操控包含统一的模式。理想状态下,图元几何元素及其基本操作应实现为编程语言的数据类型和运算符的扩展。

4.1.3 方案汇总

综上所述,可针对计算机图形学制订两种方案,即欧几里得几何体和投影几何体。

对于环境空间的描述来说，欧几里得几何体是一种较为自然的选择方案，并可以是3D 或 2D 几何体。然而，在欧几里得几何体中，转换过程并未包含统一的表达形式。除此之外，欧几里得几何体难以与几何体中的投影概念协同工作。

投影几何体则解决了欧几里得几何体的局限性。欧几里得空间包含于投影空间内，因而可使用其自然结构。另外，投影转换使得每个欧几里得转换均有一个统一的表达方式，且仍然包含投影。

本书将采用欧几里得和投影几何体这两种方案解决图形学问题，同时还将探讨两种几何体间的兼容性，进而针对各种问题使用最为适宜的处理方案。

4.2 欧几里得空间

本节将讨论欧几里得空间的属性及其元素和基本操作。

4.2.1 定义

欧几里得空间表示为维度为 $\mathbb{R}^n$ 的向量空间，并包含一个内积和内部坐标系。这一类属性也使得线性代数工具可与欧几里得空间协同工作。

此处主要关注 $\mathbb{R}^2$ 和 $\mathbb{R}^3$，即维度为 2 和 3 的欧几里得空间。

4.2.2 元素和操作

欧几里得空间 $\mathbb{R}^3$ 的基本元素是一个向量 $v = (x, y, z)$，并通过相对于标准基的坐标加以表示。

下列代码定义了 Vector3 类型和构造函数 v3_make。

```
56 <vector3 56>≡
  typedef struct Vector3 {
    Real x,y,z;
  } Vector3;
Defines:
  Vector3, used in chunks 57-59, 64, and 71-73.

57a <v3 constructor 57a>≡
  Vector3 v3_make(Real x, Real y, Real z)
  {
```

第 4 章 几 何 体

```
        Vector3 v;
        v.x = x; v.y = y; v.z = z;
        return v;
    }
Defines:
    v3_make, used in chunks 64b and 71.
Uses Vector3 56.
```

欧几里得空间的原点表示为一个零向量(0, 0, 0)。

```
57b <origin 57b>≡
    Vector3 v3_zero = { 0.0, 0.0, 0.0 };
Defines:
    v3_zero, never used.
Uses Vector3 56.
```

需要注意的是,在欧几里得空间中,向量表示为一个相对于原点的点。当表示相对于任意点的向量时,可执行坐标系变换,如图 4.1 所示。

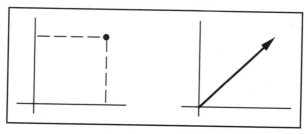

图 4.1　向量的绝对和相对含义

$\mathbb{R}^3$ 中的基本操作为向量间的加法运算和向量与标量间的乘法计算。

```
57c <v3 add 57c>≡
    Vector3 v3_add(Vector3 a, Vector3 b)
    {
        a.x += b.x; a.y += b.y; a.z += b.z;
        return a;
    }
Defines:
    v3_add, used in chunk 72c.
Uses Vector3 56.

58a <v3 scale 58a>≡
    Vector3 v3_scale(Real t, Vector3 v)
    {
```

```
    v.x *= t; v.y *= t; v.z *= t;
    return v;
  }
Defines:
  v3_scale, used in chunks 59a and 72c.
Uses Vector3 56.
```

注意，利用上述两个基本运算，还可执行诸如向量减法等其他计算。

```
#define v3_sub(a,b) v3_add(a, v3_scale(-1.0, v))
```

4.2.3 度量属性

内积在 $\mathbb{R}^3$ 中定义了一个度量，进而可计算多个重要的属性。相应地，两个向量间的内积操作通过例程 **v3_dot** 加以实现。

```
58b  <v3 dot 58b>≡
  Real v3_dot(Vector3 u, Vector3 v)
  {
    return (u.x * v.x + u.y * v.y + u.z * v.z);
  }
Defines:
  v3_dot, used in chunk 58c.
Uses Vector3 56.
```

根据内积，可计算向量的长度（或模），以及两点间的距离。

```
58c  <v3 norm alt 58c>≡
  Real v3_norm(Vector3 v)
  {
    return sqrt(v3_dot(v, v));
  }
Defines:
  v3_norm, used in chunk 59a.
Uses v3_dot 58b and Vector3 56.

#define v3_dist(a, b) v3_norm(v3_sub(a, b))
```

两个向量间角度的余弦值由向量内积和模间乘积的商定义。

```
double v3_angle(Vector3 u, Vector3 v)
{
  if (REL_EQ(0.0, v3_norm(u) * v3_norm(v)))
```

第4章 几何体

```
    error("(v3_angle) null vector");
  else
    return acos(v3_dot(u, v)/(v3_norm(u) * v3_norm(v)));
}
```

如果两个向量的内积为 0，则二者间彼此正交，也就是说，两个向量间的夹角为 90°。一个非零向量除以其模将包含单位长度。标准化向量对于表示空间方向十分有用。

```
59a <v3 unit 59a>≡
  Vector3 v3_unit(Vector3 u)
  {
    Real length = v3_norm(u);
    if(fabs(length) < EPS)
      error("(g3_unit) zero norm\n");
    else
      return v3_scale(1.0/length, u);
  }
Defines:
  v3_unit, never used.
Uses v3_norm 58c, v3_scale 58a, and Vector3 56.
```

4.2.4 坐标和基

欧几里得空间中包含了一个由标准基 $\{e_1, e_2, e_3\}$ 定义的坐标系。

```
59b <canonical basis 59b>≡
  Vector3    v3_e1    = {1.0, 0.0, 0.0};
  Vector3    v3_e2    = {0.0, 1.0, 0.0};
  Vector3    v3_e3    = {0.0, 0.0, 1.0};
Defines:
  v3_e1, never used.
  v3_e2, never used.
  v3_e3, never used.
Uses Vector3 56.
```

$\mathbb{R}^3$ 中的其他坐标系可以用 3 个线性无关向量构成的基来构造，其中较为重要的是正交基。正交基由正交的单位向量构成。

叉积也是一项特别有用的操作，特别是对于 $\mathbb{R}^3$ 中基的构造。

```
59c <v3 cross 59c>≡
  Vector3 v3_cross(Vector3 u, Vector3 v)
  {
```

```
    Vector3 uxv;
    uxv.x =   u.y * v.z - v.y * u.z;
    uxv.y = - u.x * v.z + v.x * u.z;
    uxv.z =   u.x * v.y - u.y * v.x;
    return uxv;
  }
Defines:
  v3_cross, never used.
Uses Vector3 56.
```

上述操作生成一个垂直于平面的向量，该平面由两个向量定义，其大小由这些向量构成的平行四边形的面积给出，如图 4.2 所示。

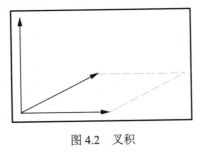

图 4.2　叉积

4.3　欧几里得空间中的转换

本节将考查欧几里得空间中的多个转换类。

4.3.1　线性转换

$\mathbb{R}^3$ 中的线性转换表示为一个运算符 $T: \mathbb{R}^3 \rightarrow \mathbb{R}^3$，并包含下列属性：

$$T(u + v) = T(u) + T(v)$$
$$T(\lambda v) = \lambda T(v)$$

其中，$u, v \in \mathbb{R}^3$，且 $\lambda \in \mathbb{R}$。

该转换类包含了多个较为理想的特性：它们保持了线性结构，并将子空间映射到子空间。一些较为重要的线性转换示例包括缩放、旋转、反射和剪切。

从计算角度来看，$\mathbb{R}^3$ 中的线性转换可表示为一个 3×3 的矩阵 M。这意味着，可利用矩阵实现转换操作。当对某个向量执行线性转换时，其操作等价于关联矩阵与向量间的乘积。

4.3.2 等距

另一类重要的转换类是等距（isometries），并保留了度量的属性，即 $\|Tv\|=\|v\|$。

欧几里得空间中的等距包括旋转、反射和平移。需要注意的是，除了平移之外，等距可视为线性转换的特例，但矩阵表达并不包含平移。

4.3.3 仿射转换

如前所述，线性转换和等距对于计算机图形学来说均十分有用，这一事实引导我们寻找一个更加通用的类，即仿射转换。仿射转换包含了空间中的线性转换和平移，同时也保持了比率和比例。

仿射转换形如 $A(x) = M(x) + v$，其中，M 表示一个矩阵，v 表示一个向量。注意，如果 A 为非线性（除非 $v = 0$），则无法通过矩阵表示 A。

对于计算机图形学中的几何体建模，仿射转换包含了较为理想的特征。该转换类从物理领域中引入了某些较为自然的几何体概念，如一致性和相似性。

但是，仿射转换也包含了一些缺点。仿射转换不支持基于矩阵的统一表达；其次，仿射转换无法实现特定的基本视图操作。例如，3D 场景照片保持平行的直线，如图 4.3 所示。因此，该操作无法通过仿射转换予以实现，仿射转换将保持不变的平行关系。

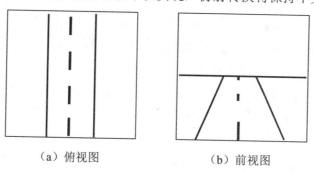

（a）俯视图　　　　（b）前视图

图 4.3　高速公路

4.4　投　影　空　间

如前所述，欧几里得空间中支持反射变换。进一步讲，如果希望针对此类操作包含统一的表达方式，并向其中引入视图转换，那么，利用投影几何体可实现这两个目标。

4.4.1 投影空间模型

维度为 n 的投影空间 $\mathbb{RP}^n$ 表示为 $\mathbb{R}^{n+1}$ 中的直线集合（穿越原点但不包括原点）。投影点 $p \in \mathbb{RP}^n$ 表示为一个等价类 $p = (\lambda x_1, \lambda x_2, ..., \lambda x_{n+1})$，且有 $\lambda \neq 0$。换而言之，$p = (x_1, x_2, ..., x_{n+1}) \equiv \lambda p$。

注意，可将维度 n 的投影空间与维度 $n+1$ 的欧几里得空间相关联。特别地，有 $\mathbb{RP}^n := \mathbb{R}^{n+1} - \{(0, 0, ..., 0)\}$。图 4.4 显示了 $\mathbb{RP}^2$ 模式，即维度 2 的投影空间，同时还显示了投影点（虚线），并省略了原点。

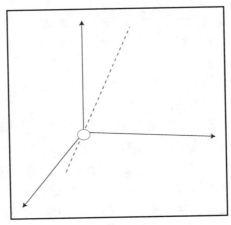

图 4.4 维度 2 的投影空间

投影空间 $\mathbb{RP}^n$ 可划分为两个集合：嵌入 $\mathbb{RP}^n$ 中的仿射空间 $\pi \equiv \mathbb{R}^n$，表示为包含 $x_{n+1} = 1$ 的投影点集合，以及包含 $x_{n+1} = 0$ 的投影点集合。

需要注意的是，根据等价关系，若 $x_{n+1} \neq 0$，可将嵌入的仿射子空间中的 n 元组（$\mathbb{R}^n$ 中的点）关联至投影点（$\mathbb{R}^{n+1}$ 中的直线）。通过这种方式，可得到下列自然分区：

$$\mathbb{RP}^n = \{(x_1, ..., x_n, x_{n+1}), x_{n+1} \neq 0\} \cup \{(x_1, ..., x_n, 0)\}$$

4.4.2 标准化和齐次坐标

根据坐标 x_{n+1} 的值，投影空间的分解可确定 $\mathbb{RP}^n$ 的特定元素集 $p = \lambda p = (\lambda x_1, \lambda x_2, \lambda x_{n+1})$，因而有：

- 仿射点：$p \in \pi$，形如 $p_a = (x_1, ..., x_n, 1)$，且有 $x_{n+1} \neq 0$，$\lambda = \dfrac{1}{x_{n+1}}$。
- 理想点：$p \notin \pi$，形如 $p_i = (x_1, ..., x_n, 0)$，且有 $x_{n+1} = 0$，$\lambda = 1$。

第 4 章 几 何 体

图 4.5 显示了嵌入的仿射平面（$z=1$）和理想平面（$z=0$）中，维度 2 的投影空间 $\mathbb{RP}^2$ 的分解过程。

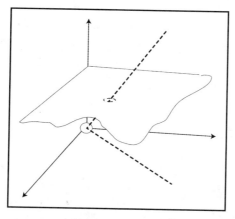

图 4.5 投影平面的分解

这里应仔细研究上述分解过程，进而与标准化坐标协同工作。实际上，可将欧几里得点视为投影点（$x_{n+1} = 1$）。尽管如此，在一般情况下，还需要与齐次坐标协同工作，形如 $p = (x_1, ..., x_n, x_{n+1})$，且无须区分仿射点和理想点。

4.4.3 齐次表达

投影空间 $\mathbb{RP}^3$ 中的一点可通过数据结构 Vector4（对应的构造函数为 v4_make）表示为齐次坐标。

```
63a <vector4 63a>≡
  typedef struct Vector4 {
    double x,y,z,w;
  } Vector4;
Defines:
  Vector4, used in chunks 63, 64, 66, and 71.

63b <v4 constructor 63b>≡
  Vector4 v4_make(Real x, Real y, Real z, Real w)
  {
    Vector4 v;
    v.x = x; v.y = y; v.z = z, v.w = w;
    return v;
  }
```

```
Defines:
  v4_make, used in chunks 64a, 66b, and 71b.
Uses Vector4 63a.
```

除此之外，还可针对仿射点和投影点间的转换定义相关例程。在此过程中，标准化操作不可或缺。注意，无法将理想点转换为一个仿射点。

```
64a <v4v3 conv 64a>≡
  Vector4 v4_v3conv(Vector3 v)
  {
    return v4_make(v.x, v.y, v.z, 1.0);
  }
Defines:
  v4_v3conv, used in chunk 71a.
Uses v4_make 63b, Vector3 56, and Vector4 63a.

64b <v3v4 conv 64b>≡
  Vector3 v3_v4conv(Vector4 v)
  {
    if (REL_EQ(v.w, 0.)) v.w = 1;
    return v3_make(v.x/v.w, v.y/v.w, v.z/v.w);
  }
Defines:
  v3_v4conv, never used.
Uses v3_make 57a, Vector3 56, and Vector4 63a.
```

4.5 $\mathbb{RP}^3$ 中的投影转换

$\mathbb{RP}^3$ 中的投影转换 T 表示为 $\mathbb{R}^4$ 中的线性运算符，如下所示。

$$T: \mathbb{R}^4 \to \mathbb{R}^4$$

通过这种方式，T 表示为一个 4×4 矩阵 M。投影转换可表示为：

$$T(p) = Mp$$

注意，当 $\lambda \neq 0$ 时，$T(p) = \lambda T(p)$。这也是投影转换和欧几里得转换间的基本差异。为了更好地理解投影转换，下面将对关联矩阵加以剖析，进而识别 4 个不同的矩阵块。

$$M = \begin{pmatrix} A & T \\ P & S \end{pmatrix}$$

❑ A：线性矩阵块（3×3）。

- T：平移矩阵块（3×1）。
- P：投影矩阵块（1×3）。
- S：缩放矩阵块（1×1）。

矩阵块 A 和 T 对应于 $\mathbb{R}^3$ 中的仿射转换，因而插入的欧几里得空间 π 保持不变。矩阵块 P 将仿射点映射至理想点，且 π 发生变化。矩阵块 S 为冗余项，其原因在于，如果 $s\neq 0$，则可假设 $s=1$，一旦 $T(p) \equiv \lambda T(p)$。

最终，我们得到了投影转换。实际上，此处整合了平移转换 $T(p) = (x + cw, y + fw, z + gw, w)$，并形成了矩阵表达形式。需要注意的是，平移的工作方式类似于与方向 w 垂直的剪切操作，如图 4.6 所示。

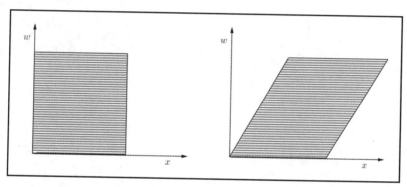

图 4.6 投影空间中的偏移转换

除此之外，投影转换还可针对 $T(p) = (x, y, z, gx + hy + iz)$ 实现透视转换。注意，该转换将仿射点转换为理想点，反之亦然。最终，理想点 $p \notin \pi$ 映射至仿射点 $p' \in \pi$，也称作消失点。也就是说，平行直线在转换后相交于另一点 p'，如图 4.7 所示。

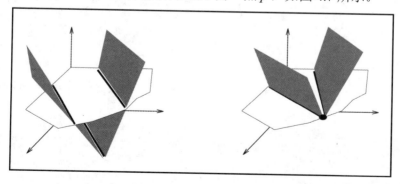

图 4.7 透视转换

投影转换将采用 4×4 矩阵表示，对此，可定义 Matrix4 类型。

```
66a <matrix4 66a>≡
  typedef struct Matrix4 {
    Vector4 r1, r2, r3, r4;
  } Matrix4;
Defines:
  Matrix4, used in chunks 66, 67, and 69-72.
Uses Vector4 63a.
```

针对向量 v 的投影转换，可通过关联矩阵 M 与 v 间的乘积实现，该计算由 v 与 M 各行的内积组成。对此，可定义一个 v4_dot 辅助例程。

```
66b <v4m4 mult 66b>≡
  Vector4 v4_m4mult(Vector4 w, Matrix4 m)
  {
    return v4_make(v4_dot(w, m.r1)
             , v4_dot(w, m.r2)
             , v4_dot(w, m.r3)
             , v4_dot(w, m.r4));
  }
Defines:
  v4_m4mult, never used.
Uses Matrix4 66a, v4_dot 66c, v4_make 63b, and Vector4 63a.

66c <v4 dot 66c>≡
  Real v4_dot(u,v)
       Vector4 u, v;
  {
    return (u.x * v.x + u.y * v.y + u.z * v.z + u.w * v.w);
  }
Defines:
  v4_dot, used in chunks 66b, 69, and 71.
Uses Vector4 63a.
```

接下来将考查基本转换的实现过程。其中，第一个转换是恒等转换 m4_ident，如下所示。

```
66d <m4 ident 66d>≡
  Matrix4 m4_ident()
  {
    Matrix4 m = {{1.0, 0.0, 0.0, 0.0},
                 {0.0, 1.0, 0.0, 0.0},
```

```
                  {0.0, 0.0, 1.0, 0.0},
                  {0.0, 0.0, 0.0, 1.0}};
    return m;
  }
Defines:
  m4_ident, used in chunk 67.
Uses Matrix4 66a.
```

其中，平移转换如下所示。

$$M_t = \begin{pmatrix} 1 & 0 & 0 & t_x \\ 0 & 1 & 0 & t_y \\ 0 & 0 & 1 & t_z \\ 0 & 0 & 0 & 1 \end{pmatrix}$$

并通过例程 m4_translate 予以实现，如下所示。

```
67a <m4 translate 67a>≡
  Matrix4 m4_translate(Real tx, Real ty, Real tz)
  {
    Matrix4 m = m4_ident();
    m.r1.w = tx;
    m.r2.w = ty;
    m.r3.w = tz;
    return m;
  }
Defines:
  m4_translate, never used.
Uses m4_ident 66d and Matrix4 66a.
```

沿主轴方向上的缩放转换如下所示。

$$M_s = \begin{pmatrix} s_x & 0 & 0 & 0 \\ 0 & s_y & 0 & 0 \\ 0 & 0 & s_z & 0 \\ 0 & 0 & 0 & 1 \end{pmatrix}$$

并通过例程 m4_ scale 予以实现，如下所示。

```
67b <m4 scale 67b>≡
  Matrix4 m4_scale(Real sx, Real sy, Real sz)
  {
    Matrix4 m = m4_ident();
    m.r1.x= sx;
    m.r2.y= sy;
```

```
      m.r3.z= sz;
      return(m);
   }
Defines:
   m4_scale, never used.
Uses m4_ident 66d and Matrix4 66a.
```

例程 m4_rotate 在 $\mathbb{R}^3$ 内通过欧拉角在主轴方向上实现了旋转转换，如下所示。

```
67c <m4 rotate 67c>≡
   Matrix4 m4_rotate(char axis, Real angle)
   {
      Matrix4 m = m4_ident();
      Real cost = (Real) cos(angle);
      Real sint = (Real) sin(angle);
      switch (axis) {
      case 'x' :
         m.r2.y= cost; m.r2.z= -sint;
         m.r3.y= sint; m.r3.z= cost;
         break;
      case 'y' :
         m.r1.x= cost; m.r1.z= sint;
         m.r3.x= -sint; m.r3.z= cost;
         break;
      case 'z' :
         m.r1.x= cost; m.r1.y= -sint;
         m.r2.x= sint; m.r2.y= cost;
         break;
      default :
         error("(m4_rotate) invalid axis\n");
      }
      return m;
   }
Defines:
   m4_rotate, never used.
Uses m4_ident 66d and Matrix4 66a.
```

截至目前，我们得到了欧几里得空间中的基本转换。除此之外，这里还将实现其他有用的转换操作，如反射和剪切。除了欧几里得转换操作，透视转换在视图处理过程中也十分重要（第 11 章还将对此加以讨论），如下所示。

$$M_p = \begin{pmatrix} 1 & 0 & 0 & 0 \\ 0 & 1 & 0 & 0 \\ 0 & 0 & 1 & 0 \\ m & n & p & 1 \end{pmatrix}$$

矩阵代数与投影变换代数之间存在着一种自然的同构关系。通过这种方式，转换的组合等价于矩阵的连接操作，如下所示。

$$p' = T_n (\ldots T_2 (T_1 (p)))$$
$$p' = M_n \ldots M_2 M_1 p$$
$$p' = M_p$$

实际上，可以用一个矩阵来表示由任意转换序列产生的变换，这也体现了统一矩阵表达方式的优点。

需要注意的是，矩阵的连接操作不符合交换律，这反映了以下事实：转换序列的结果取决于每个转换的操作顺序，如下所示。

$$M_1 M_2 \ldots M_n \neq M_n \ldots M_2 M_1$$

转换序列的逆可通过下列方式获得：逆序连接的、矩阵的逆连接，如下所示。

$$(M_1 M_2 \ldots M_n)^{-1} = M_n^{-1} \ldots M_2^{-1} M_1^{-1}$$

例程 **m4_m4prod** 实现了矩阵的乘积，这也是转换组合的基本操作。

```
69 <m4m4 prod 69>≡
 Matrix4 m4_m4prod(Matrix4 a, Matrix4 b)
 {
   Matrix4 m, c = m4_transpose(b);

   m.r1.x = v4_dot(a.r1, c.r1);
   m.r1.y = v4_dot(a.r1, c.r2);
   m.r1.z = v4_dot(a.r1, c.r3);
   m.r1.w = v4_dot(a.r1, c.r4);

   m.r2.x = v4_dot(a.r2, c.r1);
   m.r2.y = v4_dot(a.r2, c.r2);
   m.r2.z = v4_dot(a.r2, c.r3);
   m.r2.w = v4_dot(a.r2, c.r4);

   m.r3.x = v4_dot(a.r3, c.r1);
   m.r3.y = v4_dot(a.r3, c.r2);
   m.r3.z = v4_dot(a.r3, c.r3);
   m.r3.w = v4_dot(a.r3, c.r4);
```

```
    m.r4.x = v4_dot(a.r4, c.r1);
    m.r4.y = v4_dot(a.r4, c.r2);
    m.r4.z = v4_dot(a.r4, c.r3);
    m.r4.w = v4_dot(a.r4, c.r4);
    return m;
  }
Defines:
  m4_m4prod, never used.
Uses Matrix4 66a and v4_dot 66c.
```

出于简单性考虑，例程 m4_m4prod 使用了输入矩阵之一的转置结果。注意，这并非是实现矩阵间乘积的高效方式。

例程 m4_transpose 实现了矩阵的转置操作，对于矩阵通用计算来说，这是一项十分有用的操作。

```
70 <m4 transpose 70> ≡
  Matrix4 m4_transpose(Matrix4 m)
  {
    Matrix4 mt;
    mt.r1.x= m.r1.x;
    mt.r1.y= m.r2.x;
    mt.r1.z= m.r3.x;
    mt.r1.w= m.r4.x;

    mt.r2.x= m.r1.y;
    mt.r2.y= m.r2.y;
    mt.r2.z= m.r3.y;
    mt.r2.w= m.r4.y;

    mt.r3.x= m.r1.z;
    mt.r3.y= m.r2.z;
    mt.r3.z= m.r3.z;
    mt.r3.w= m.r4.z;

    mt.r4.x= m.r1.w;
    mt.r4.y= m.r2.w;
    mt.r4.z= m.r3.w;
    mt.r4.w= m.r4.w;
    return mt ;
  }
Uses Matrix4 66a.
```

4.6 几何体对象的转换

针对图形对象表达中的多种元素，本节讨论如何对其应用转换操作。

4.6.1 转换操作修正

这里，所转换的基本元素是一个向量 $p = (x, y, z) \in \pi \equiv \mathbb{RP}^3$，且隶属于嵌入的仿射空间 $\pi \in \mathbb{RP}^3$，因而将使用标准化的表达方式 $p = (x, y, z, 1)$。

当对齐次向量 p 应用投影矩阵 M 所示的转换时，可执行操作 $p' = Mp$。需要注意的是，如果在转换后希望维持向量的标准化形式，则需要执行所谓的坐标齐次除法，即除以分量 w。该操作实际上对应于齐次向量在嵌入的仿射空间内的投影。

$$p' = \frac{1}{w'}(x', y', z', w')$$
$$p'' = (x'', y'', z'', 1)$$

4.6.2 转换点和方向

例程 v4_m4mult 将齐次向量乘以投影矩阵，该例程用于通用的投影转换。

例程 v3_m4mult 将欧几里得向量乘以一个投影矩阵，该例程用于仿射转换。

```
71a <v3m4 mult 71a>≡
  Vector3 v3_m4mult(Vector3 v, Matrix4 m)
  {
    Vector4 w = v4_v3conv(v);
    return v3_make(v4_dot(w, m.r1), v4_dot(w, m.r2), v4_dot(w, m.r3));
  }
Defines:
 v3_m4mult, used in chunk 72b.
Uses Matrix4 66a, v3_make 57a, v4_dot 66c, v4_v3conv 64a, Vector3 56, and
Vector4 63a.
```

例程 v3_m3mult 将向量乘以投影矩阵的线性转换部分。该例程对于转换方向向量十分有用，且不会受到平移转换的影响。

```
71b <v3m3 mult 71b>≡
  Vector3 v3_m3mult(Vector3 v, Matrix4 m)
  {
```

```
    Vector4 w = v4_make(v.x, v.y, v.z, 0.0);
    return v3_make(v4_dot(w, m.r1), v4_dot(w, m.r2), v4_dot(w, m.r3));
  }
```
Defines:
  v3_m3mult, used in chunk 72b.
Uses Matrix4 66a, v3_make 57a, v4_dot 66c, v4_make 63b, Vector3 56, and Vector4 63a.

4.6.3 转换射线

许多视图操作常会涉及光学模拟。其中，基本的几何对象表示为射线 r，并通过原点 o 和方向 d 加以定义，如图 4.8 所示。

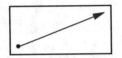

图 4.8 表示射线的向量

相应地，此处将定义一个 Ray 类型，对应的构造函数为 ray_make。

```
typedef struct Ray {
  Vector3 o, d;
} Ray;

72a <ray constructor 72a>≡
  Ray ray_make(Vector3 o, Vector3 d)
  {
    Ray r;
    r.o = o; r.d = d;
    return r;
  }
```
Defines:
  ray_make, never used.
Uses Ray and Vector3 56.

射线 r 的基本操作包括应用仿射转换，并根据参数 t 计算射线上的一点，如 $pt = o + td$。

```
72b <ray transform 72b>≡
  Ray ray_transform(Ray r, Matrix4 m)
  {
    r.o = v3_m4mult(r.o, m);
    r.d = v3_m3mult(r.d, m);
```

        return r;
    }
Defines:
  ray_transform, never used.
Uses Matrix4 66a, Ray, v3_m3mult 71b, and v3_m4mult 71a.

72c ⟨ray point 72c⟩≡
  Vector3 ray_point(Ray r, Real t)
  {
    return v3_add(r.o, v3_scale(t, r.d));
  }
Defines:
  ray_point, never used.
Uses Ray, v3_add 57c, v3_scale 58a, and Vector3 56.

在一些视见问题中，常有必要计算射线与表面间的交点。对此，可使用数据结构 Inode，其中包含了必要的信息，如对应于交点的参数 t，在交点处垂直于表面的向量 n，等等。该结构可用作一个链表元素，其中包含了沿射线上的所有相交信息。

72d ⟨inode 72d⟩≡
  typedef struct Inode {
    struct Inode   *next;
    double   t;
    Vector3   n;
    int   enter;
    struct Material *m;
  } Inode;
Defines:
  Inode, used in chunk 73.
Uses Vector3 56.

考虑到信息类型变化较大，因而此处定义了构造函数 inode_alloc 和析构函数 inode_free。

73a ⟨inode constructor 73a⟩≡
  Inode *inode_alloc(Real t, Vector3 n, int enter)
  {
    Inode *i = NEWSTRUCT(Inode);
    i->t = t;
    i->n = n;
    i->enter = enter;
    i->next = (Inode *)0;
    return i;
  }

```
Defines:
  inode_alloc, never used.
Uses Inode 72d and Vector3 56.

73b <inode destructor 73b>≡
  void inode_free(Inode *l)
  {
    Inode *i;
    while (l) {
      i = l; l = l->next; free(i);
    }
  }
Defines:
  inode_free, never used.
Uses Inode 72d.
```

4.6.4 切平面上的转换

与点和向量相比，某点处表面切平面的转换稍有不同。

通过对应于平面隐式方程的系数，可采用行向量 $n = (a, b, c, d)$ 表示一个平面 n，因此，每个点 $p \in n$ 满足下列方程：

$$\{p = (x, y, z, 1) \mid ax + by + cz + d = 0\}$$

该方程可利用内积 $\langle n, p \rangle = 0$ 并通过余弦方式加以构造。

如果通过矩阵 M 向平面 n 应用转换，在经由 M 转换后，n 中的点 p 满足 $(nM^{-1})(Mp) = 0$ 这一条件。也就是说，转换后的点 Mp 位于转换平面 nM^{-1} 上。

在该示例中可以看到，当利用列向量转换切平面时，应采用逆矩阵的转置结果，如下所示。

$$n' = (M^{-1})^T n$$

注意，对于正交矩阵，有 $(M^{-1})^T = M$。这是唯一一种可以把切平面转换成向量的情况。

4.6.5 转换的双重解释

转换操作可以理解为向量的转换或者是坐标系之间的变化。

第一种解释将转换视为同一坐标系中点间的映射。通过这种方式，点 p 映射为 $T(p)$。这种解释方式有助于理解图形对象的参数描述。其中，几何体表示为一个函数 $p=g(u, v)$，并定义了该对象上的点。当转换参数对象时，可向此类点直接应用转换矩阵，即 $p' = T(g(u, v))$，如图 4.9 所示。

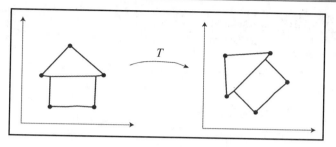

图 4.9　直接转换：映射点

第二种解释则将转换视为坐标系的变化。通过这一方式，标准基 $\{e_1, e_2, e_3\}$ 中的向量 v 将被映射为对应于转换基 $v'= xT(e_1) + yT(e_2) + zT(e_3)$ 的向量。

该解释有助于理解图形对象的隐式描述，其中，几何体通过环境空间 $v \in \mathbb{R}^3$ 的函数 $h(v) = 0$ 加以定义。当转换隐式对象时，可向转换空间的点应用逆转换，这将在原始环境空间中进行计算，即 $h(T^{-1}(v)) = 0$，如图 4.10 所示。

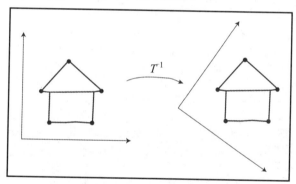

图 4.10　逆转换：坐标系的变化

4.7　补 充 材 料

针对计算机图形学中的基本几何操作，本章讨论了相关库的实现过程。

Jim Clark 发表了首篇与视见转换相关的论文 The Geometry Engine[Clark 82]，他同时也是 Silicon Graphics 创始人。该论文描述了一个实现几何转换的图形处理器。

4.7.1　小结

GEOM 库的外部 API 由下列例程构成：

```
Vector3 v3_make(Real x, Real y, Real z);
Vector3 v3_scale(Real t, Vector3 v);
Vector3 v3_add(Vector3 a, Vector3 b);
Vector3 v3_sub(Vector3 a, Vector3 b);
Vector3 v3_cross(Vector3 u, Vector3 v);
Vector3 v3_unit(Vector3 u);
Real v3_dot(Vector3 u, Vector3 v);
Real v3_norm(Vector3 v);

Vector4 v4_make(Real x, Real y, Real z, Real w);
Vector3 v3_v4conv(Vector4 w);
Vector4 v4_v3conv(Vector3 v);

Vector3 v3_m4mult(Vector3 v, Matrix4 m);
Vector3 v3_m3mult(Vector3 v, Matrix4 m);
Vector4 v4_m4mult(Vector4 w, Matrix4 m);

Matrix4 m4_ident();
Matrix4 m4_translate(Real tx, Real ty, Real tz);
Matrix4 m4_scale(Real sx, Real sy, Real sz);
Matrix4 m4_rotate(char axis, Real angle);
Matrix4 m4_transpose(Matrix4 m);
Matrix4 m4_m4prod(Matrix4 a, Matrix4 b);

Ray ray_make(Vector3 o, Vector3 d);
Ray ray_transform(Ray r, Matrix4 m);
Vector3 ray_point(Ray r, Real t);

Inode *inode_alloc(Real t, Vector3 n, int enter);
void inode_free(Inode *l);
```

4.7.2 程序设计层

针对 3D 计算机图形学，GEOM 库实现了基本的几何操作。其中，Vector3 和 Matrix4 涵盖了图形程序中所用的大多数数据结构。这里，几何转换直接实现于图形卡硬件中（需要支持 OpenGL 和 DirectX 标准）。

4.8 本章练习

（1）使用 GEOM 库实现 vexpr 程序，并计算向量表达式。其中，操作数表示为 x、

y、z 格式的 3D 向量,同时,运算符集合应至少包含向量加法、向量与标量的乘法、内积、叉积和向量间的相等测试。

当前程序需要从 stdin 中读取运算数之一,其他运算数则从命令行中读取。相关运算操作应通过名称指定,并作为命令行中的参数被传递(如 add、mult),最终结果将发送至 stdout 中,例如:

```
echo 1 3 0.4 | vexpr add 2 1 0.1
3 4 0.5
```

(2)使用 GP 和 GEOM 库,实现交互式程序并转换和绘制多边形直线。相关转换操作须包括平移、旋转和缩放。

(3)扩展 GEOM 库并包含下列功能:
① 两个向量间的线性插值。
① 向量 u 在向量 v 上的正交投影。
③ 向量 u 在向量 v 上的切线投影。
④ 向向量中添加一个分量。
⑤ 针对既定向量,计算垂直的单位向量。
⑥ 计算 3 点构成的平面的正交向量。
⑦ 计算一个正交转换矩阵,并将既定单位向量 u 映射至既定的单位向量 v 上。
⑧ 将 4×4 矩阵元素置于 Matrix4 数据结构中(使用 C 语言)。
⑨ 计算沿任意轴向上的旋转矩阵。
⑩ 计算 4×4 矩阵的逆矩阵。

第5章 颜　　色

颜色是计算机图形学中较为重要的元素之一，常用于感受物理环境的视觉系统中。另外，颜色信息也是图像的主要属性。除此之外，光照模拟也会涉及颜色计算。本章将讨论颜色及其在计算机图形学中的应用。

5.1 颜色的基本知识

本节将使用之前讨论的 4 种环境范例研究计算机图形学中的颜色。通过这种方式，首先考查物理环境下的颜色，随后定义颜色的数学模型、构建颜色的表达方式，进而生成计算机设备中颜色的实现结构。

颜色是由电磁辐射通过人体视觉系统所引起的感觉。因此，颜色是一种物理学现象。这一事实具有很大的相关性，因而研究颜色时既要考虑感性方面，又要考虑物理方面的内容。

5.1.1 颜色的波长模型

从物理学角度来看，颜色源自电磁辐射，其波长 λ 位于光谱的可见波段，约为 400～800 纳米，如图 5.1 和 5.2 所示。

图 5.1　可见光谱中的颜色

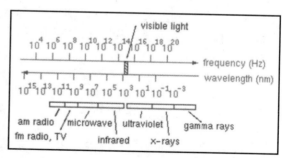

图 5.2　电磁光谱中的波段

颜色的感知是由视觉系统接收到的多个波长的电磁辐射组合而成的。颜色特征通过其光谱分布函数定义，进而将能量值与每个波长 λ 关联，如图 5.3 所示。

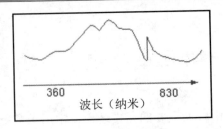

图 5.3　光谱分布函数

基于光谱分布的颜色特征使我们可以得到如下结论：函数是描述颜色的数学模型。更为准确地讲，一种给定的颜色可表示为 $D = \{f : [\lambda_0, \lambda_1] \in \mathbb{R}^+ \to \mathbb{R}^+\}$ 光谱分布空间中的一个元素。其中，$[\lambda_0, \lambda_1]$ 表示为定义可见光谱波段的区间。需要注意的是，光谱分布空间定义为一个函数空间，因而包含了无限的维度。

5.1.2　物理颜色系统

在物理环境中，颜色通过物理颜色系统被处理。这一类系统通常划分为颜色接收系统和颜色发射系统。相关示例包括相机（接收系统）和监视器（发射系统）。

颜色接收系统由一组传感器 $\{s_i\}, i = 1, \dots, n$ 构成，并在可见光谱的 n 度中进行采样。每个传感器的响应结果由 $c_i = \int c(\lambda) s_i(\lambda) d\lambda$ 确定。通过这种方式，颜色接收系统将每个光谱分布与样本向量相关联，即 $c\{\lambda\} = (c_1, \dots, c_n)$。

颜色发射系统由一组发射器 $\{e_i\}, i = 1, \dots, n$ 构成，进而生成包含特定光谱分布 $e_i = d_i(\lambda)$ 的电磁能量。由该系统发出的颜色通过每个发射器分布函数的线性组合加以确定，即 $c(\lambda) = \sum c_i d_i(\lambda)$。

注意，物理颜色系统包含有限维度，并体现了一种颜色光谱离散化的自然方法。接收系统执行采样，而发射系统执行重构操作。据此，可将光谱分布函数转换为采样，反之亦然。相应地，与接收器和发射器关联的光谱分布函数称作系统的主颜色。采样向量 $(c_1, \dots c_n)$ 提供了系统的颜色表达结果。

5.1.3　色彩的心理学研究

可以看到，物理颜色系统需要转化为光谱分布的连续模型。因此，我们得到了一个与整体物理颜色系统相关联的有限维度空间。

当理解颜色的感知内容时，我们也应研究一下人类的视觉系统。人类的眼睛可看作是一个物理颜色接收系统，其中包含了对应于低（红色）、高（绿色）和高（蓝色）波长波段的 3 种传感器类型。相关试验表明，视觉系统的颜色空间涵盖了潜在的线性结构。

第 5 章 颜 色

这一结果得到了 Grassmans 定律的验证，其中，颜色匹配遵循线性和相加性[Malacara-Hernandez 02]。该定律是由德国学者赫尔曼·冈瑟·格拉斯曼（Hermann Gunther Grassmann，1809—1877）发现的。

感知色彩的一个直接结果是，可以使用三维向量空间作为视觉系统的数学模型。更为准确地讲，可使用欧几里得空间 $\mathbb{R}^3$，其中，基向量 $\{e_1, e_2, e_3\}$ 与系统的主颜色相关联。

类似地，也可为其他物理颜色系统定义颜色空间。此项任务也被 CIE（国际照明委员会）所实现，并定义了多个标准的颜色系统，如 CIE-RGB 和 CIE-XYZ 系统。

上述事实表明，可使用欧几里得几何体工具与颜色空间协同工作。相应地，可通过一个 3D 向量表示为某种颜色，其坐标对应于系统主颜色的分量。接下来，我们将根据主颜色红、绿、蓝采用 RGB 系统。

```
79 <color 79>≡
  typedef Vector3 Color;

  #define RED(c) (c.x)
  #define GRN(c) (c.y)
  #define BLU(c) (c.z)
Defines:
BLU, used in chunks 83b and 88a.
Color, used in chunks 83b and 86-88.
GRN, used in chunks 83b and 88a.
RED, used in chunks 83b and 88a.
```

这里定义了一个颜色构造函数 c_make、颜色加法操作 c_add、颜色与标量的乘积 c_scale，以及颜色乘法 c_mult。这一类运算各自包含了相应的物理含义：求和计算对应于颜色的合并；缩放操作表示为提升颜色的亮度；两种颜色向量间的乘积则等价于过滤机制。

另外一方面，除了颜色乘积之外，此类操作对于向量空间来说也较为自然。

```
#define c_make(r,g,b) v3_make(r,g,b)
#define c_add(a, b)   v3_add(a,b)
#define c_scale(a, b) v3_scale(a, b)
#define c_mult(a, b)  v3_mult(a, b)
```

两种颜色间的乘积如下所示：

```
80a <v3 mult 80a>≡
  Vector3 v3_mult(Vector3 a, Vector3 b)
  {
    Vector3 c;
    c.x = a.x * b.x; c.y = a.y * b.y; c.z = a.z * b.z;
```

```
    return c;
  }
Defines:
  v3_mult, never used.
```

5.1.4 颜色计算

上述例程定义了颜色的基本操作。

此外，还可通过矩阵与向量间的乘法运算 v3_m4mult 修改基向量。这一类转换类型对应于不同颜色的物理系统间的转换（维度为 3 的线性颜色空间）。

```
80b <colconvert 80b> ≡
  Vector3 col_convert(Vector3 c, Matrix m)
  {
    return v3_m4mult(c, m);
  }
Defines:
  col_convert, never used.
```

需要注意的是，当计算执行两个物理系统 A 和 B 间转换的矩阵 m_{AB} 时，需要使用到与两个系统主颜色相关联的评估分布函数。

当计算图形时，大多数问题可通过三原色空间加以解决。然而，某些较为特殊的问题则需要通过复杂的计算得以实现，包括非线性转换，甚至是基于重采样的光谱分布函数的重构。

第一种情形涉及包含不同维度或非线性空间的颜色空间的转换；第二种情况出现于与波长相关的光照现象中。图 5.4 显示了各种颜色计算过程。

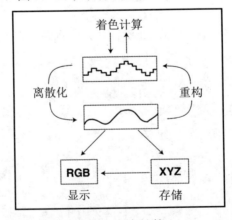

图 5.4 颜色计算

5.2 设备颜色系统

在物理颜色系统中，颜色的显示坐标一般为正值，且有所限制，其原因在于传感器和发射器就可操作的电磁能量来说包含一定的物理限制，因此，一种较为方便的方法是将颜色实体与设备的物理系统关联，该实体针对设备定义了一个有效的颜色集。

5.2.1 颜色的处理

物理颜色系统中，另一个较为重要的问题是颜色格式的处理。颜色格式的处理与主颜色的光谱分布函数合成方式相关，该过程可以是颜色的加法或减法。在加法处理过程中，合成操作通过光谱分布的叠加（加法）完成；在减法处理过程中，合成操作根据白色通过对光谱分布进行滤波（乘法）完成。

监视器和视频投影仪的 mRGB 系统即是加法系统中的示例。mRGB 采用主颜色（红、蓝、绿）的混合机制。区间[0,1]内的标准化颜色坐标将转换为单位立方体中的颜色立方体，如图 5.5 所示。对于加法系统来说，原点对应于黑色。

图 5.5　RGB 颜色立方体

```
#define C_WHITE c_make(1,1,1)
#define C_BLACK c_make(0,0,0)
```

对于颜色处理格式为减法的颜色系统，相关示例包括彩色打印机的 CMY 系统和摄影机制。在 CMY 系统中，青色、洋红色和黄色的油墨将对白色进行过滤。CMY 系统中的颜色实体同时也是一个单位立方体。在该系统中，原点对应于白色。

5.2.2 RGB-CMY 转换

RGB 和 CMY 系统间的转换十分简单，其中仅涉及某些基本的变化。在每个主方向上通过 1 进行缩放将执行加法和减法间的转换。系统的原点则通过沿立方体对角线(1,1,1)上的平移操作被映射，如图 5.6 所示。

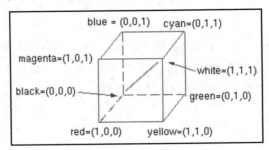

图 5.6 CMYK 颜色立方体

matrix rgb_cmy_m 例程实现了上述转换。

```
82a <rgb cmy matrix 82a> ≡
  Matrix4 rgb_cmy_m = {{-1.0,  0.0,  0.0, 1.0},
                       { 0.0, -1.0,  0.0, 1.0},
                       { 0.0,  0.0, -1.0, 1.0},
                       { 0.0,  0.0,  0.0, 0.0}
  };
Defines:
  rgb_cmy_m, used in chunks 82b and 83a.
```

转换过程实际上通过 rgb_to_cmy 和 cmy_to_rgb 例程完成，其中使用了相同的矩阵。

```
82b <rgb to cmy 82b> ≡
  Vector3 rgb_to_cmy(Vector3 c)
  {
    return v3_m4mult(c, rgb_cmy_m);
  }
Defines:
  rgb_to_cmy, never used.
Uses rgb_cmy_m 82a.

83a <cmy to rgb 83a> ≡
  Vector3 cmy_to_rgb(Vector3 c)
  {
    return v3_m4mult(c, rgb_cmy_m);
  }
Defines:
  cmy_to_rgb, never used.
Uses rgb_cmy_m 82a.
```

5.3 颜色规范系统

颜色规范系统旨在直观地识别不同的颜色。相应地，考查感知特征是实现这一目标的关键。这里的问题是，人类识别颜色的重要参数是什么？在回答了这一问题后，即可针对颜色规范定义相应的颜色空间。

5.3.1 亮度：色度分解

从直观上看，颜色多多少少会包含一定的量度，这对应于与光谱分布所关联的能量的变化。当给定光谱分布函数 $c(\lambda)$ 和一个实数 $t > 0$，乘积 $c' = tc(\lambda)$ 对应于另一个光谱分布，其中仅对波长 λ 包含了非 0 能量值，且 $c(\lambda)$ 也为非 0。如果 $t > 0$，c' 则包含较大的能量值；如果 $0 < t < 1$，c' 则包含较小的能量值。从感知角度来看，函数 $c(\lambda)$ 的缩放过程将改变亮度所主导的量度，或者颜色的亮度。

颜色的亮度由运算符 $L(c) = \sum l_i c_i$ 定义，其中，l_i 取决于颜色系统的主颜色。在监视器的 mRGB 系统中，$l_r = 0.299$，$l_g = 0.587$，$l_b = 0.114$。

```
83b <rgb to y 83b> ≡
  Real c_rgb_to_y(Color c)
  {
    return 0.299 * RED(c) + 0.587 * GRN(c) + 0.114 * BLU(c)
  }
Defines:
  c_rgb_to_y, never used.
Uses BLU 79, Color 79, GRN 79, and RED 79.
```

颜色信息可划分为两个分量，即亮度和色度。其中，亮度与颜色的光亮程度相关；而色度则与彩色程度有关，也就是说，与颜色无关的密度变化。

在维度 3 的颜色空间内，颜色 $c \in \mathbb{R}^3$ 可通过其坐标 $c = (c_1, c_2, c_3)$ 表示。如前所述，当 $t > 0$ 时，向量 $tc = (tc_1, tc_2, tc_3)$ 表示包含可变亮度的同一色度信息。随后可得出，色度空间表示为一个投影空间。

为了简化色度的表达，可定义一个子集 3，其中，每个点对应于一个唯一的色度信息。一种较为适宜的选取法是定义一个 $c_1 + c_2 + c_3 = 1$ 平面，在该平面中，空间主颜色组合生成的每种颜色均包含一个表达结果。该平面称作 Maxwell 平面。

在 Maxwell 平面上，颜色的放射投影坐标称作色度坐标。此类坐标的计算相对简单。包含同一色度的和不同亮度的两种颜色，在 Maxwell 平面上仅存在一种表达结果 c_i'，其坐标满足 $c_1' + c_2' + c_3' = 1$。通过这种方式，可针对任意颜色得到 t，以使 $tc = c'$。因此，

$$t(c_1 + c_2 + c_3) = c_1' + c_2' + c_3' = 1$$

即：

$$t = \frac{1}{c_1 + c_2 + c_3}$$

$$c_i' = \frac{c_i}{c_1 + c_2 + c_3}$$

从感知角度来看，将颜色分解为亮度和色度十分重要。这意味着，可将颜色向量 c 表示为两个向量之和，即 $c = c_l + c_c$。其中，c_l 描述了颜色的亮度，而 c_c 则描述了色度信息，如图 5.7 所示。这一分解过程还将在稍后讨论颜色规范空间时再次被提及。

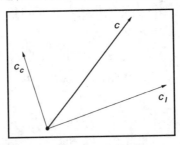

图 5.7　颜色向量 c 分解为亮度分量 c_l 和色度分量 c_c

5.3.2　颜色选择的 HSV 系统

对于颜色选择来说，亮度和色度分解可生成更加直观的坐标。该处理过程包括选择颜色亮度以及确定颜色色度。

色度可通过系统空间色度中的一点加以选择。由于色度集具有 2D 特性，因而可在平面上构建极坐标系统。在该系统中，原点对应于白色。当远离原点时，将会得到更多的饱和色或纯色；而原点的径向方向将会得到不同的着色颜色。

从感知角度来看，HSV 系统即是基于此类较为直观的参数的（色调、饱和度或相对于亮度的数值）。HSV 空间可以自然地与一个六角形基面的金字塔相关联，其顶点位于原点处。在颜色实体中，对应值在 0～1 变化，即金字塔的顶端至底部。在与金字塔轴向正交的平面上，可利用极坐标描述着色和饱和度，如图 5.8 所示。

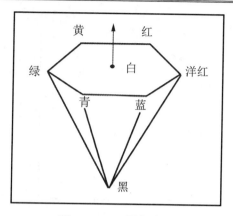

图 5.8　HSV 颜色实体

RGB-HSV 转换。当执行 RGB 和 HSV 系统间的转换时，需要在两种系统的颜色实体间进行映射。鉴于 RGB 立方体和 HSV 金字塔顶点间存在自然的对应关系，因而该项任务得到了简化。

此处将坐标值 v 和颜色 $c = (r, g, b)$ 定义为 $v(c) = \max(r, g, b)$。通过这种方式，针对每个值 v，可得到与单位 RGB 立方体平行的立方体 C_v，如图 5.9 所示。倒金字塔的顶点对应于黑色(0, 0, 0)，中心位置则对应于白色(1, 1, 1)。因此，可得到 HSV 金字塔轴线与 RGB 立方体对角线之间的对应关系。

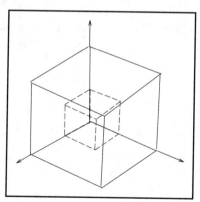

图 5.9　平行于单位 RGB 立方体的立方体

在垂直于对角线的平面 π_v 上生成立方体 C_v 的正交投影并通过点 (v, v, v)，可得到一个六边形，其中，每个顶点对应于某个主 RGB 颜色或互补色。也就是说，RGB 立方体的剩余顶点将被映射至 HSV 金字塔六边形基面的顶点，如图 5.10 所示。

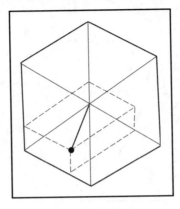

图 5.10 六边形

例程 rgb_to_hsv 负责 RGB 系统和 HSV 系统间的转换操作。颜色值计算参照定义 $v = \max(r, g, b)$ 进行。

饱和度计算则需要确定平面 π_v 上向量 $c = (r, g, b)$ 的正交投影 c_c 与边界间的分式，简而言之，即标准化 $s \in [0, 1]$。

$$s = \frac{v - \min(r, g, b)}{v}$$

色调计算需要确定投影向量 c 和向量 R 间的角度。需要注意的是，取决于 c 的最小分量，其投影 c_c 位于一对六边形的三角形扇形中。更加准确地讲，令 $x = \min(r, g, b)$，如果 $x = r$，则有 $c_c \in \{(R, Y, W) \cup (Y, G, W)\}$；如果 $x = g$，则有 $c_c \in \{(R, M, W) \cup (M, B, W)\}$；如果 $x = b$，则有 $c_c \in \{(G, C, W) \cup (C, B, W)\}$。在此类扇形中，某个 CMY 顶点的相对坐标如下所示。

$$h = \frac{a - b}{v - x}$$

其中，a 和 b 分别表示为扇形对初始顶点和最终顶点的对应分量。

```
86 <rgb to hsv 86>≡
  Color rgb_to_hsv(Real r, Real g, Real b)
  {
    Real v, x, f;
    int i;

    x = MIN(r, MIN(g, b));
    v = MAX(r, MAX(g, b));
```

```
    if (v == x)
      return v3_make(UNDEFINED, 0, v);
    f = (r == x) ? g - b : ((g == x) ? b - r : r - g);
    i = (r == x) ? 3 : ((g == x) ? 5 : 1);
    return c_make(i - f /(v - x), (v - x)/v, v);

  }
Defines:
  rgb_to_hsv, never used.
Uses c_make and Color 79.
```

例程 hsv_to_rgb 负责将 HSV 系统转换为 RGB 系统。

```
87 <hsv to rgb 87>≡
  Color hsv_to_rgb(Real h, Real s, Real v)
  {
    Real m, n, f;
    int i;

    if(h == UNDEFINED)
      return c_make(v, v, v);
    i = floor(h);
    f = h - i;
    if(EVEN(i))
      f = 1 - f;
    m = v * (1 - s);
    n = v * (1 - s * f);
    switch (i) {
    case 6:
    case 0: return c_make(v, n, m);
    case 1: return c_make(n, v, m);
    case 2: return c_make(m, v, n);
    case 3: return c_make(m, n, v);
    case 4: return c_make(n, m, v);
    case 5: return c_make(v, m, n);
    }
  }
Defines:
  hsv_to_rgb, never used.
Uses c_make and Color 79.
```

5.4　离散化颜色实体

图形设备中的颜色通过整数加以表示，这意味着，在实际操作过程中，图形设备与离散化的颜色空间协同工作。总体而言，设备中的颜色信息通过一个向量定义，其分量为包含 n 位精确度的整数。

注意，颜色空间离散化对应于设备颜色实体的均匀区域。在第 6 章介绍数字图像颜色离散化问题时，还将继续讨论如何构建颜色实体的非均匀区域。

某些图形设备使用了 m 位的整数表示颜色信息。对此，有必要定义一种方法，进而包装和展开该表达方式中的颜色分量。

```
88a <rgb to index 88a>≡
  int rgb_to_index(Color c, int nr, int ng, int nb)
  {
    unsigned int r = CLAMP(RED(c), 0, 255);
    unsigned int g = CLAMP(GRN(c), 0, 255);
    unsigned int b = CLAMP(BLU(c), 0, 255);

    r = (r >> (8 - nr)) & MASK_BITS(nr);
    g = (g >> (8 - ng)) & MASK_BITS(ng);
    b = (b >> (8 - nb)) & MASK_BITS(nb);

    return ((r << (ng + nb))| (g << nb)| b);
  }
Defines:
  rgb_to_index, never used.
Uses BLU 79, Color 79, GRN 79, MASK_BITS 88b, and RED 79.

88b <index to rgb 88b>≡
  Color index_to_rgb(int k, int nr, int ng, int nb)
  {
    unsigned int r, g, b;
    r = ((k >> (ng + nb)) & MASK_BITS(nr)) << (8 - nr);
    g = ((k >> (nb)) & MASK_BITS(ng)) << (8 - ng);
    b = ((k) & MASK_BITS(nb)) << (8 - nb);

    return c_make(r, g, b);
  }
```

```
#define MASK_BITS(n) ((01 << (n))-1)
Defines:
  index_to_rgb, never used.
  MASK_BITS, used in chunk 88a.
Uses c_make and Color 79.
```

5.5 补充材料

本章讨论了颜色的表达方式及其库的实现。针对颜色选择和转换，图 5.11 显示了两个示例。

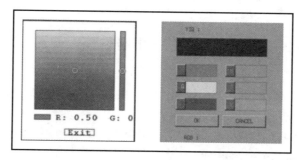

图 5.11 颜色的选择和转换

在许多图形对象中，颜色是较为重要的属性之一。在图形设备中，库例程常用于不同颜色表达方式间的转换，以及用户的颜色规范中。

Alvy Ray Smith 发表了首篇与 HSV 系统相关的论文[Smith 81]，并于近期发表了一篇名为 HWB: A More Intuitive Hue-Based Color Model 的论文[Smith and Lyons 96]。

5.5.1 资料链接

关于颜色标准，CIE 和 ICC 是两家重要的国际机构，读者可访问下列链接以了解更多信息。

- CIE：国际照明委员会，对应网址为 http://www.hike.te.chiba-u.ac.jp/ikeda/CIE/home.html。
- ICC：国际色彩联盟，对应网址为 http://www.color.org/。

另外，Charles Poyton 的个人网站中也包含了一些与颜色相关的有趣信息，对应网址为 http://www.inforamp.net/~poynton/notes/colour/and/gamma/ColorFAQ.html。

5.5.2 回顾

颜色库中的 API 包含了以下例程:

```
Vector3 rgb_to_cmy(Vector3 c);
Vector3 cmy_to_rgb(Vector3 c);
Vector3 rgb_to_yiq(Vector3 c);
Vector3 yiq_to_rgb(Vector3 c);

Vector3 rgb_to_hsv(Real r, Real g, Real b);
Vector3 hsv_to_rgb(Real h, Real s, Real v);
```

5.6 本章练习

（1）编写程序，构建包含 256 种颜色的颜色表。针对 RGB 颜色实体，该程序应至少实现两种离散化方法。

（2）利用 GP 库编写程序，显示练习（1）中生成的颜色表，并对两种方法进行比较。

（3）设计并实现颜色选择微件。

（4）当给定任意一个 RGB 颜色和颜色图，编写一个函数，从该颜色图中找到完美表示它的颜色。

（5）编写一个程序，并在 RGB 和 HSV 系统间转换颜色。

第 6 章 数 字 图 像

图像作为视觉过程的最终结果,在计算机图形学中具有重要意义,它们是交互系统不可缺少的一部分,因此在建模过程中也扮演着重要的角色。本章将讨论数字图像、表达方式以及图像操作。

6.1 基 础 知 识

对于图像这一概念来说,我们需要再次考查之前所讨论的 4 种环境范例,并在物理环境中分析图像的特征,对其定义数学模型,构建图像的表现模式,并针对图像编码制定数据格式,如图 6.1 所示。

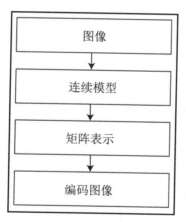

图 6.1 图像的抽象级别

6.1.1 图像的离散和连续模型

图像由 2D 支撑构成,且每个点关联相应的颜色信息。通过这种方式,可将函数 $f: U \subset \mathbb{R}^2 \to C$ 用作图像的数学模型。其中,集合 U 表示为图像支撑,f 的数值集合称作图像色域。在该模型中,图像的函数域通常表示为一个矩形 $U = [a, b] \times [c, d]$,对应的反域表示为三色空间 $C = \mathbb{R}^3$,如 RGB 空间。

当在计算机上表达一幅图像时，需要对图像函数的域和反域离散化。相应地，采样是几何体支撑的一种离散化方案；量化方案则是颜色空间的一种离散化方法，或者是降低色域的一种处理方法，如图 6.2 所示。

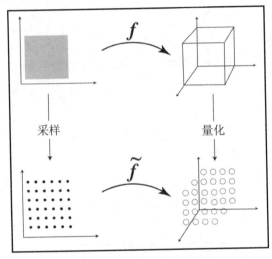

图 6.2　图像的离散化

6.1.2　图像的量化

量化过程可视为一种转换，即 $q: \mathbb{R}^n \to M_k$。其中，$M_k = \{c_1, c_2, \dots c_k\}$ 定义为 $\mathbb{R}^n$ 的有限子集。这里，集合 M_k 称作量化转换的颜色图。若 $k = 2^l$，则称 q 为 l 位的量化结果。当给定图像的离散表达方式后，在该类型的有限颜色子集之间进行量化是很常见的操作，即 $q: R_j \to M_k$。如果 $j = 2^n$，且 $k = 2^m$，则可得到 n 到 m 位的量化结果。

考查量化转换 $q: \mathbb{R}^n \to M_k$。M_k 的元素 c_i 称作量化级别。在每个量化级别上，$c_i \in M_k$ 对应于颜色 $C_i \subset \mathbb{R}^n$ 的一个子集，表示通过转换 q 映射至 c_i 上的颜色，如下所示。

$$C_i = q^{-1}(c_i) = \{c \in C; q(c) = c_i\}$$

集合 c_i 族构成了颜色空间划分，该划分的每个 C_i 集合称作量化单元。需要注意的是，量化函数 q 假设每个单元 C_i 中的常量等于 c_i。通过使用几何参数，可以得到始于 C_i 的 c_i，反之亦然。通过这种方式，一些量化方法可首先在 c_i 级别上执行计算，并于随后在单元 C_i 处进行计算，而其他则使用相反的策略。

如果 q 通过量化单元 C_i 给出，级别 c_i 对应于单元的形心位置。如果 q 通过量化级别 c_i 确定，单元 C_i 的几何体则对应于与级别 c_i 关联的 Voronoi 图。换而言之，每个单元 C_i 由最靠近级别 c_i 的颜色空间中的点集形成。

6.1.3 矩阵表达

图像支撑的采样通常根据均匀网格 $F_\Delta = \{(u_i, v_j) \in \mathbb{R}^2\}$ 予以执行。其中，$u_i = a + i\Delta u$，$v_j = c + j\Delta v$，且 $\Delta u = (b - a)/m$，$\Delta v = (d - c)/n$，$i = 0, ... m$，$j = 0, ... n$。相应地，点 (u_i, v_j) 称作采样点。

通过 2D 矩阵，可表示一幅离散图像。在矩阵表达形式中，可将图像函数 $f(u, v)$ 与矩阵 $A_{m \times n} = \{a_{ij}\}$ 关联。其中，$i = 1, m$；$j = 1, n$。

离散图像元素 $a_{ij} = f(u_i, v_j)$ 称作像素（源自"图像元素"）。在单色图像中，$a_{i,j} = x$ 表示为一个标量；而在三色图像中，$a_{i,j} = (r, g, b)$ 表示为一个向量。而且，$a_{i,j} = k$ 可表示为颜色图中对应颜色 $c_k = (r, g, b)$ 的索引。

图像的可见分辨率则通过采样数量或者是矩阵 A 的行数 m 和列数 n 加以确定。图像颜色分辨率则通过颜色级别量或 a_{ij} 的位数确定。

6.2　图像的表现格式

综上所述，本节将针对图像表达构建相应的格式。

6.2.1　数据结构

这里将采用矩阵方式作为图像格式。当实现这一格式时，将涉及两种信息类型，如下所示。

- 头：定义了表达数据，如空间分辨率和量化类型。
- 像素矩阵：图像矩阵的元素。其中，像素表示为 RGB 颜色向量。

数据结构 Image 由包含空间分辨率的头，以及指向像素矩阵的指针构成。

```
94a <image structure 94a>≡
  typedef struct Image {
    int w, h;
    unsigned int maxval;
    Bcolor *c;
  } Image;
Defines:
  Image, used in chunks 94-97.
Uses Bcolor 94b.
```

图像元素由 8 位整数定义。另外，像素表示为 3D 向量 Bcolor，其中，每个分量表示为一个 Byte。

```
94b <image elements 94b>≡
  typedef unsigned char Byte;

  typedef struct Bcolor {
    Byte z, y, x;
  } Bcolor;
Defines:
  Bcolor, used in chunk 94.
  Byte, never used.
```

除此之外，还通过宏 PIX_MIN 和 PIX_MAX 并针对 Byte 定义了一个有效值区间。

```
94c <pixelvalues 94c>≡
  #define PIX_MIN 0
  #define PIX_MAX 255
Defines:
  PIX_MAX, used in chunk 95c.
  PIX_MIN, used in chunk 95c.
```

例程 img_init 表示为结构 Image 的构造函数，其参数为空间分辨率和图像类型。

```
94d <imginit 94d>≡
  Image *img_init(int type, int w, int h)
  {
    Image *i = NEWSTRUCT(Image);

    i->w = w; i->h = h;
    i->maxval = 255;
    i->c = NEWTARRAY(w*h, Bcolor);
    img_clear(i, C_BLACK);
    return i;
  }
Defines:
  img_init, used in chunk 97.
Uses Bcolor 94b, Image 94a, and img_clear
```

例程 img_free 定义为结构 Image 的析构函数。

```
95a <imgfree 95a> ≡
  void img_free(Image *i)
```

```
    {
      efree(i->c); efree(i);
    }
Defines:
  img_free, never used.
Uses Image 94a.
```

6.2.2 访问图像矩阵

此处选择在连续的内存空间中存储图像元素。据此，可通过宏 PIXRED、PIXGRN 和 PIXBLU 访问图像矩阵中的元素(u,v)。

```
95b <image array access 95b>≡
  #define PIXRED(I,U,V) I->c[U + (((I->h - 1) - (V)) * I->w)].x
  #define PIXGRN(I,U,V) I->c[U + (((I->h - 1) - (V)) * I->w)].y
  #define PIXBLU(I,U,V) I->c[U + (((I->h - 1) - (V)) * I->w)].z
Defines:
  PIXBLU, used in chunks 95c and 96a.
  PIXGRN, used in chunks 95c and 96a.
  PIXRED, used in chunks 95c and 96a.
```

例程 img_putc 和 img_getc 可通过颜色访问图像矩阵。

```
95c <imgputc 95c> ≡
  void img_putc(Image *i, int u, int v, Color c)
  {
    if (u >= 0 && u < i->w && v >= 0 && v < i->h) {
      PIXRED(i,u,v) = CLAMP(RED(c), PIX_MIN, PIX_MAX);
      PIXGRN(i,u,v) = CLAMP(GRN(c), PIX_MIN, PIX_MAX);
      PIXBLU(i,u,v) = CLAMP(BLU(c), PIX_MIN, PIX_MAX);
    }
  }
Defines:
  img_putc, used in chunk 97.
Uses Image 94a, PIX_MAX 94c, PIX_MIN 94c, PIXBLU 95b, PIXGRN 95b, and
  PIXRED 95b.

96a <imggetc 96a>≡
  Color img_getc(Image *i, int u, int v)
  {
    if (u >= 0 && u < i->w && v >= 0 && v < i->h)
```

```
      return c_make(PIXRED(i,u,v),PIXGRN(i,u,v),PIXBLU(i,u,v));
    else
      return C_BLACK;
  }
Defines:
  img_getc, used in chunk 96b.
Uses Image 94a, PIXBLU 95b, PIXGRN 95b, and PIXRED 95b.
```

6.3 图像编码

对于图像表示，此处选择了 PPM 格式（可移植的像素图），大多数 Unix 系统都支持这种格式，Windows 中也有一些商业和公共领域的程序支持这种格式。

6.3.1 PPM 格式

PPM 格式并不复杂，并支持矩阵表达中所用的图像类型。PPM 文件一般以 .ppm 扩展名结尾，并包含了头和像素矩阵。其中，头包含了与矩阵表达相关的常用信息，如空间分辨率。该文件的前两个字节对应于整数 PPM_MAGIC，进而可验证文件是否为 PPM 格式。

```
#define PPM_MAGIC P6
```

6.3.2 直接编码

如前所述，PPM 文件由头和像素矩阵构成。图像矩阵的直接编码是通过按行排序的矩阵元素表得到的。读取和写入 PPM 图像则采用了开源库 libnetpbm（对应网址为 http://netpbm.sourceforge.net）。

例程 img_write 将图像数据写入 PPM 文件中。

```
96b <imgwrite 96b>≡
  void img_write(Image *i, char *fname, int cflag)
  {
    FILE *fp;
    int row, col;
    pixval maxval;
    pixel **pixels;
```

```
  fp = (strcmp("stdout", fname) == 0)? stdout : fopen(fname, "wb");
  pixels = ppm_allocarray(i->w, i->h);
  for ( row = 0; row < i->h; ++row ) {
    for ( col = 0; col < i->w; ++col ) {
      Color c = img_getc(i, col, i->h - row - 1);
      pixel p = pixels[row] [col] ;
      (pixels[row] [col]).r = RED(c);
      (pixels[row] [col]).g = GRN(c);
      (pixels[row] [col]).b = BLU(c);
    }
  }
  ppm_writeppm(fp, pixels, i->w, i->h, i->maxval, TRUE);
  pnm_freearray(pixels, i->h);
  if (strcmp("stdout", fname) ! = 0)
    fclose(fp);
}
Defines:
  img_write, never used.
Uses Image 94a and img_getc 96a.
```

例程 img_read 将 PPM 文件读取至 Image 结构中。

```
97 <imgread 97>≡
  Image *img_read(char *fname)
  {
    FILE *fp, *fp2;
    int row, col, cols, rows;
    pixval maxval;
    pixel **pixels;
    Image *i;

    fp = (strcmp("stdin", fname) == 0)? stdin : fopen(fname, "rb");
    pixels = ppm_readppm(fp, &cols, &rows, &maxval);
    i = img_init(0, cols, rows);
    i->maxval = maxval;
    for ( row = 0; row < rows; ++row ) {
      for ( col = 0; col < cols; ++col ) {
        pixel const p = pixels[row] [col] ;
        img_putc(i, col, rows-row-1, c_make(PPM_GETR(p), PPM_GETG(p),
              PPM_GETB(p)));
      }
    }
```

```
    pnm_freearray(pixels, rows);
    fclose(fp);
    return i;
  }
Defines:
 img_read, never used.
Uses Image 94a, img_init 94d, and img_putc 95c.
```

6.4 补充材料

本章讨论了图像的表达方式,并使用了相关库操控图像数据。图 6.3 显示了图像可视化的程序示例。关于图像处理和计算图形学,读者还可参考 [Gomes and Velho 95, Gomes and Velho 97]。

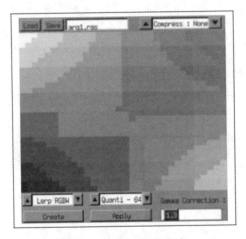

图 6.3 图像可视化程序

6.4.1 修正

图像库 API 包含了下列例程:

```
Image *img_init(int type, int w, int h);
void img_clear(Image *i, Color c);
Image *img_read(char *fname);
void img_write(Image *i, char *fname, int cflag);
```

```
void img_putc(Image *i, int u, int v, Color c);
Color img_getc(Image *i, int u, int v);

void img_free(Image *i);
```

6.4.2 图像格式

一些较为重要的图像格式包括：
- TIFF。
- GIF。
- JPEG。
- PhotoCD。

一些用于图像的公共域包包括：
- Utah Raster Toolkit。
- PBM。
- LibTIFF。
- IRIS Tools。

GIMP、ImageMagic 和 XV 则是浏览图像时较为流行的公共域程序。

6.5 本章练习

（1）编写程序，生成包含以下模式的图像：
① 正方形。
② 灰色度的渐变效果。
③ 在图像角点处 4 种颜色的插值计算结果。
④ 白噪声。

（2）编写程序，分别实现下列功能：
① 使用公共域程序（如 GIMP）可视化练习（1）中的图像。
② 利用 GP 库可视化图像。
③ 对①和②的结果进行比较，并讨论程序的局限性。

（3）编写程序执行图像上的 Gamma 修正，借助该程序，并根据经验确定监视器的 Gamma 系数。

（4）编写一个抖动程序，采用练习（1）中的灰色调模式，并针对 6、4、2、1 位进行量化测试。

（5）编写一个量化程序，利用练习（1）中创建的模式，并针对 6、4、2、1 位进行量化测试。

（6）编写一个程序并利用 RLE 方法压缩图像，利用练习（1）中生成的图像测试该程序。针对每幅图像验证压缩因子，并对结果进行解释。

（7）编写一个程序，并组合颜色量化和抖动方法。

（8）编写一个程序，将彩色图像转换为灰色度图像。

第 7 章　3D 场景描述

与数字图像一样，三维场景也是计算机图形学的基本要素之一。视见处理始于三维场景的描述，最终生成一幅数字图像；建模处理过程在用户的指导下生成某个场景的表达结果。本章将讨论计算方法的表达内容，进而构建三维场景。

7.1　三维场景

三维场景描述了计算机图形学应用程序中的虚拟世界，该表现方式的主要功能是捕捉虚拟世界中与建模和可视化相关的内容。

场景描述构建了建模处理和可视化处理间的界面，如图 7.1 所示。除此之外，这种表现方式也用于场景的存储和传输方面，如 CAD 系统或分布式虚拟现实系统。

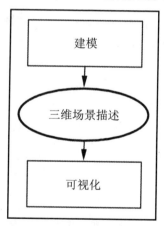

图 7.1　建模处理和可视化处理间的界面

7.1.1　三维场景的元素

三维场景描述由其中的元素规范以及构建方式构成，同时也包含了配置信息。三维场景组件表示为虚拟世界中的对象，此类对象可分为 3 类，如下所示。

- 对象模型：描述几何体和虚拟属性。

- 光源：描述光照。
- 虚拟相机：表述观察者。

三维场景中的对象结构对应于基于位置的对象分组，如几何链接。此类分组可通过层次结构方式加以构建。另外，三维场景中的配置可通过多个参数予以确定，进而指定了对描述内容的控制方式，如场景名称、场景图像的数据等。

需要注意的是，三维场景描述涉及多种信息类型和参数。这种多样性也体现了三维场景描述中需要考查的基本特征。

7.1.2　三维场景表达

类似于数字图像格式的定义（参见第 6 章），也可对三维场景执行相同的操作。然而，仅对此加以定义还远远不够，三维场景中包含了多种信息类型，并在特定的场景中予以表现（也有可能未曾出现）。最终，设置一种固定的数据格式则较为复杂、低效。

另一种建议则具有一定的灵活性，即根据通用编程语言采用过程式表达方式，如 C 语言或 Lisp 语言。但这需要使用到缺乏色场景语义的上下文环境结构，因而该方案也难以令人满意。

针对事务场景的表述，较为理想的方法处于上述两种极端方案之间，在特殊性和普遍性的基础上兼具灵活性和强大的表现力。该方法使用了场景描述语言。

7.1.3　场景描述语言

描述三维场景的语言应满足下列要求：
- 直观的符号。
- 统一的语法。
- 可扩展的语义。
- 简单的实现。

除此之外，该语言还应可从现有的领域标准中引入相关概念，其中较为重要的是 Open Inventor SDL（场景描述语言）和 VRML（虚拟现实建模语言）。基于这两个标准，即可开发一种三维场景描述语言，进而满足上述需求。该语言可通过简单而高效的构建方式描述场景元素。

【例 7.1】　简单的三维场景。

```
scene {
    camera = view { from = {0, 0, -2.5}, fov = 90},
```

```
light = ambi_light { intensity = 0.2 },
light = dist_light { direction = {0, 1, -1} },
object = group{
        material = plastic { kd = 0.8, ks = 0.0 },
        transform = { translate {v = {0, .0, 0}}},
        children = {
           primobj{ shape = sphere{radius = .1 }},
           primobj{ shape = cylinder { center = {2, 2, 2}}}}},
}
```

上述示例描述了由虚拟相机、两个光源和包含两个图元对象的分组构成的三维场景。注意，此类信息以清晰、准确的方式进行编码，同时表达了规则的结构。

下面将讨论与编程语言相关的计算概念。根据这一类概念，将确定三维场景描述语言的实现，并展示其具体应用。

7.2 语言概念

语言是描述信息的系统图式。语言的语法决定其形式结构，而语言的语义与语言的内容有关。表达式由一组根据语法规则构造的符号组成，其内容由语义定义予以确定。在编程语言中，内容表示为一个计算过程。

7.2.1 表达式语言

鉴于存在多种编程语言类型，这里将重点关注表达式语言。表达式语言基于两种基本的抽象概念，即运算符和操作数。

在这一类型的语言中，程序由顺序的表达式集合构成。对应语言则包含了相关规则，进而构建表达式及其评估语义。注意，由于评估是一种计算处理过程，因而相关内容出现于流程的执行过程中。

表达式语言与函数计算密切相关。实际上，可将运算符与函数关联，并将操作数与参数关联。在表达式语言中，每个表达式均包含了一个值。表达式可以是原始表达式，或者是组合表达式。其中，原始表达式包含直接定义的值（也就是说，针对语言基本类型的既定常量）。组合表达式的值由集合的递归评估结果定义（即函数及其参数的应用）。

这里，我们将采用替代模型实现表达式的评估，基本过程涵盖了以下结构：

```
if simple expression
    returns value
```

```
if composed expression
    finds values of the operands
    returns the value of the applied operator to the operands
```

7.2.2 表达式中的语法和语义

在表达式语言中,语法规则的目标是判断运算符和操作数以构成表达式。注意,针对相同的语义内容,可包含不同的语法内容。例如,考查由一个运算符和两个操作数构成的二元表达式。对此,存在 3 种可能构建二元表达式,这取决于运算符相对于操作数的位置,且分别对应于前缀、后缀和中缀运算符。

【例 7.2】 求和操作。
- ❑ 前缀符号:(+ a b)。
- ❑ 后缀符号:(a b +)。
- ❑ 中缀符号:(a + b)。

语法分析将生成与所用符号无关的表达式结果,并通过语法树予以显示。语法树的内部节点包含运算符;而叶节点则包含了基本类型。层次结构由子表达式的聚类构成。稍后将显示相关示例。

【例 7.3】 表达式((a + b) * (c − (d + e)))的树形结构如图 7.2 所示。该表达式采用了前缀符号。注意,同一树形结构也可用于表示后缀和前缀符号设置的表达式。

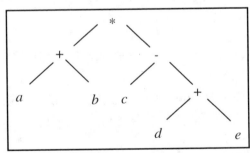

图 7.2 表达式树形结构

语法分析器标识的表达式被传递至表达式评估器(实现了 7.2.1 节中的替代模型)。通过这一方式,执行过程包含了构成当前程序的、顺序型评估表达式树形结构。下列伪代码显示了表达式语言的解释器结构。

```
While (more expressions)
    Read expression
    Evaluate expression
```

7.2.3 程序的编译和解释

程序通常采用构成文本或源代码的数字-字符进行编码。我们需要对相关文本进行分析，以便从中提取高级语言的结构，并于随后对此类结构进行处理，从而执行程序的具体内容。

其中，源代码分析阶段称作编译。在这一阶段中，将对词法和句法结构进行分析。后续处理阶段称作执行阶段，并将语义内容归属于计算结构。

在特定的计算系统中，分析和执行阶段可通过称作解释器的单一程序实现。编译器（或解释器）由下列模块构成，进而实现了分析阶段和程序代码的执行过程。

- ❑ 词法分析器（或扫描器）。
- ❑ 语法分析器（或解析器）。
- ❑ 表达式评估器（或评估器）。

除了上述 3 种模块之外，还应包含相应的符号管理模块（符号管理器）以及错误处理机制（错误处理程序）。

在词法分析阶段，字符序列经分组后形成基本的语言类型，包括数字、名称等内容。语法分析器将源代码作为输入，并将基本类型的标识符（标记）作为输出结果。相应地，名称通过符号管理器存储于一张表中。在表达式语言中，语法确定了组合表达式的构建方式，这一过程始于基本类型。语法分析器将标识符作为输入内容，并将表达式树形结构作为输出结果。

如前所述，语义通过语言运算符加以确定。表达式评估器运行当前程序，并将运算符应用于操作数上。图 7.3 显示了程序解释器的结构、模块和彼此间的相互关系。

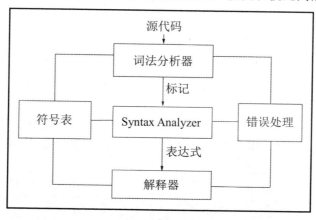

图 7.3　解释器结构

7.2.4 语言开发工具

UNIX 程序开发环境中提供了多种工具可促进语言的实现，并由 Open Software Foundation 予以改进，同时也是 GUN 编程包的一部分内容。下列工具可生成实现了解释器（或编译器）模块的程序。这些模块由所设计语言的词汇、语法和语义规则指定。

- lex：扫描器生成器（词法分析）。
- yacc：解析器生成器（语法分析）。
- CPP/M4：宏转换器（预处理）。

7.3 扩展语言

这里，我们并不打算直接实现三维场景描述语言，而是开发一种"元语言"，进而实现一种渐进式且具有一定灵活性的语言。此处将使用称为扩展语言的元语言类型，它具有封闭、统一的语法和开放、最小化的语义。

本节将采用称之为扩展语言的元语言类型，其中包含了封闭和统一的语法，以及开放和最小化的语义。这一模式的优点是，扩展语言针对语法分析整合了全部计算环境；同时还支持应用程序语义规范，如场景描述。

扩展语言针对"子语言"（或嵌入式语言）提供了计算支持。此类子语言内核的语义，以及用户定义的函数。

本节将开发包含扩展语言解释器的所有模块。

7.3.1 语法分析器

此处将使用 UNIX 工具开发扩展语言，特别地，将使用 yacc 程序生成语法分析器。作为输入，yacc 中包含了一个语法规则规范，用于构建该语言的表达式树。表达式树则由终结符和非终结符构成。

终结符（或标记）是对应于树叶节点的基本语言类型。当前语言涵盖了以下基本类型：数字、名称和类。字符串是引号之间的字符序列串，而类则是运算符的标识符。

```
107a <tokens 107a> ≡
  %token <dval> NUMBER
  %token <sval> STRING NAME
  %token <fval> CLASS
```

非终结符（或类型）表示为对应于内部树节点的、当前语言的语法结构。当前语言包含了以下非终结符：value、pv、pvlist、expression 和 input。

```
107b <types 107b>≡
  %type <nval> node input
  %type <pval> pvlist pv
  %type <vval> val
```

下列结构显示了表达式树的节点（内部节点或外部节点）。据此，数据结构由所有基本类型和派生类型的组合构成。

```
107c <parserdata structure 107c>≡
  %union {
    char ival;
    double dval;
    char *sval;
    Pval *pval;
    Val vval;
    Node *nval;
    Val (*fval)();
  }
Uses Node 115b, Pval 113b, and Val 114a.
```

通过以下语法规则，我们定义了当前语言的语法形式。

```
108 <grammarrules 108>≡
  %%
  input:  /* empty */              { $$ = root = NULL;}
        | input node ';'           { $$ = root = $2;}
        | input error ';'          { yyerrok; $$ = root = NULL;}
        ;
  node:   CLASS '{' pvlist '}'     { $$ = t_node($1, $3);}
        | '{' pvlist '}'           { $$ = t_node(t_pvl, $2);}
        ;
  pvlist: /* empty */              { $$ = (Pval *)0;}
        | pvlist ','               { $$ = $1;}
        | pvlist pv                { $$ = pv_append($1, $2);}
        ;
  pv:     NAME '=' val             { $$ = pv_make($1, $3);}
        | val                      { $$ = pv_make(NULL, $1);}
        ;
  val:    NUMBER                   { $$ = pv_value(V_NUM, $1, NULL, NULL);}
        | '-' NUMBER               { $$ = pv_value(V_NUM, - $2, NULL, NULL);}
```

```
            | STRING              { $$ = pv_value(V_STR, 0., $1, NULL);}
            | node                { $$ = pv_value(V_NOD, 0., NULL, $1);}
            ;
%%
Uses pv_append 115a, pv_make 114b, pv_value 114c, Pval 113b, root, t_node
115c, t_pvl 117a, V_NOD 113c, V_NUM 113c, and V_STR 113c.
```

上述规则定义了简单且功能强大的语法，同时支持类型的构建。

```
class { name = value, name = value }
```

这种结构非常适合描述 n 元表达式。其中，运算符表示为 class 类型；操作数则表示为 name 和 value 对，并可对其加以命名。

由于 value 也可表示为表达式，因而可对子表达式进行分组以形成表达式。进一步讲，可从基本结构中省略 class 和 name，并生成{name = value, name = value}，class{value, value}和{value, value}。需要注意的是，鉴于 value 可以是一个数字，因而最后一个构造结果可以表示 n 维向量。

一项较为重要的观察结果与错误管理有关。在第一项语法规则中，我们得到了一个形如"input error;"的结果。这意味着，当检测到语法错误时，部分数据将会被丢弃，同时例程 yyerror 将被调用。该例程简单地通知用户输入文件中的某一行处出现了错误。

```
109a <yyerror 109a>≡
  int yyerror()
  {
    extern int lineno;
    fprintf(stderr, "lang: syntax error in, near line %d\n",lineno);
  }
Defines:
  yyerror, never used.
Uses lineno 109c.
```

7.3.2 词法分析器

词法分析器负责处理源代码，将输入字符分组，进而形成符号（或标记）。对此，可使用程序 lex 生成词法分析器。然而，这里推荐直接对其予以实现，从而可对处理过程拥有更多的控制权。

该程序较为简单，其工作方式类似于有限状态机。根据当前字符，算法决定是否处于当前状态，或者调整状态。其中，每一种状态对应于一种符号类型。对应状态由 yylex 例程中宏的名称标识。

第 7 章 3D 场景描述

```
109b <yylex 109b>≡
  yylex()
  {
    Symbol *s;

    SCAN_WHITE_SPACE
    SCAN_NUMBER
    SCAN_STRING
    SCAN_NAME_CLASS
    SCAN_MARK
  }
Uses SCAN_MARK 111b, SCAN_NAME_CLASS 111a, SCAN_NUMBER 110b, SCAN_STRING
110c, SCAN_WHITE_SPACE 110a, Symbol 112b, and yylex.
```

词法分析器内部信息存储于变量 fin 中，该变量表示为指向输入文件的指针；lineno 和 c 则表示源代码中的行和字符。这些变量还可由词法分析器的其他例程访问。

```
109c <internalstate 109c>≡
  static FILE *fin = stdin;
  static int c;
  int lineno = 0;
Defines:
  c, used in chunks 110, 111, and 117a.
  fin, used in chunks 110-12.
  lineno, used in chunks 109a and 111b.
```

宏 SCAN_WHITE_SPACE 的功能是检测代码中的空格，此外还将检测条件 EOF，即文件的结尾。

```
110a <scan white space 110a>≡
  #define SCAN_WHITE_SPACE
    while ((c = getc(fin)) == ' ' || c == '\t')
      ;
    if (c == EOF) return 0;
Defines:
  SCAN_WHITE_SPACE, used in chunk 109b.
Uses c 109c and fin 109c.
```

宏 SCAN_NUMBER 构造由浮点数序列组成的数字。

```
110b <scan number 110b>≡
  #define SCAN_NUMBER
    if (c == '.' || isdigit(c)) { double d;
      ungetc(c, fin);
      fscanf(fin, "%lf", &d);
```

```
        yylval.dval = d;
        return NUMBER;
    }
```
Defines:
  SCAN_NUMBER, used in chunk 109b.
Uses c 109c and fin 109c.

宏 SCAN_STRING 构建用引号分隔的字符串。

110c <scan string 110c>≡
```
  #define SCAN_STRING
    if (c == '"') { char sbuf[320], *p;
      for (p = sbuf; (c = getc(fin)) ! = '"'; *p++ = c) {
        if (c == '\\')
          c = ((c = getc(fin)) == '\n')? ' ' : c;
        if (c == '\n'|| c == EOF) {
          fprintf(stderr,"missing quote, sbuf\n"); break;
        }
        if (p >= sbuf + sizeof(sbuf) - 1) {
          fprintf(stderr,"sbuffer overflow\n"); break;
        }
      }
      *p = '\0';
      yylval.sval = malloc(strlen(sbuf) + 1);
      strcpy(yylval.sval, sbuf);
      return STRING;
    }
```
Defines:
  SCAN_STRING, used in chunk 109b.
Uses c 109c and fin 109c.

宏 SCAN_NAME_CLASS 通过符号管理器标识一个名称,稍后还将对此加以讨论。

111a <scan name class 111a> ≡
```
  #define SCAN_NAME_CLASS
    if (isalpha(c)) { char sbuf[1024], *p = sbuf; int t;
      do {
        if (p >= sbuf + sizeof(sbuf) - 1) {
          *p = '\0';
          fprintf(stderr,"name too long %s (%x: %x)\n", sbuf, p, sbuf);
        }
        *p++ = c;
      } while ((c = getc(fin)) ! = EOF && (isalnum(c)|| c == '_' ));
      ungetc(c, fin);
```

第 7 章 3D 场景描述

```
        *p = '\0';
      if ((s = sym_lookup(sbuf)) == (Symbol *)0)
        s = sym_install(sbuf, NAME, NULL);
      if (s->token == CLASS)
        yylval.fval = s->func;
      else
        yylval.sval = s->name;
      return s->token;
    }
Defines:
  SCAN_NAME_CLASS, used in chunk 109b.
Uses c 109c, fin 109c, sym_install 112c, sym lookup 113a, and Symbol 112b.
```

宏 SCAN_MARK 用于检测字母数字和特殊字符，如行结束符（\n）。

```
111b <scan mark 111b>≡
  #define SCAN_MARK  \
    switch (c) {    \
    case '\\': if ((c = getc(fin)) != '\n')  \
                 return c;    \
               else   \
                 return yylex();  \
    case '\n': lineno++; return yylex();  \
    default: return c;   \
    }
Defines:
  SCAN_MARK, used in chunk 109b.
Uses c 109c, fin 109c, lineno 109c, and yylex.
```

例程 yyfile 用于指定词法分析器的输入文件。

```
112a <yyfile 112a>≡
  void yyfile(FILE *fd)
    {
      fin = fd;
    }
Defines:
  yyfile, never used.
Uses fin 109c.
```

7.3.3 符号分析器

符号分析器用于维护程序的名称表。数据结构 Symbol 用于表示表中的元素。这里，一个符号可以是一个名称或者是运算符的标识符。在当前示例中，表包含了一个指向实现了运算符的函数。

```
112b <symboltable 112b>≡
  typedef struct Symbol {
    char *name;
    int token;
    Val (*func)();
    struct Symbol *next;
  } Symbol;
Defines:
  Symbol, used in chunks 109b and 111-13.
Uses Val 114a.
```

字符表是一个存储于内部变量 symlist 中的链表。例程 sym_install 将在表中安装一个字符。

```
112c <symbolinstall 112c>≡
  static Symbol *symlist = (Symbol *)0;

  Symbol *sym_install(char *s, int t, Val (*func)())
  {
    Symbol *sp = (Symbol *) malloc(sizeof(Symbol));

    sp->name = malloc(strlen(s) + 1);
    strcpy(sp->name, s);
    sp->token = t;
    sp->func = func;
    sp->next = symlist;
    return symlist = sp;
  }
Defines:
  sym_install, used in chunks 111a and 118c.
  symlist, used in chunk 113a.
Uses Symbol 112b and Val 114a.
```

例程 sym_lookup 检测某个名称是否对应于表中的已有符号。若该符号已存在，则返回其指针；否则将返回一个空指针。

```
113a <symbollookup 113a>≡
  Symbol *sym_lookup(char *s)
  {
    Symbol *sp;
```

```
    for (sp = symlist; sp ! = (Symbol *)0; sp = sp->next) {
      if (strcmp(sp->name, s) == 0)
        return sp;
    }
    return (Symbol *)0;
  }
Defines:
  sym_lookup, used in chunks 111a and 118c.
Uses Symbol 112b and symlist 112c.
```

7.3.4 参数和值

语言表达式操作数由参数列表确定。相应地，一个参数由一对名称和数值构成，结构 Pval 用于表示该对。注意，这是一个单链表元素。

```
113b <parametervalue structure 113b>≡
  typedef struct Pval {
    struct Pval *next;
    char *name;
    struct Val val;
  } Pval;
Defines:
  Pval, used in chunks 107c, 108, and 114-18.
Uses Val 114a.
```

值对应于当前语言的各种基本类型和派生类型。如前所述，它们分别表示为数字、字符串、表达式和参数列表。此外，还可定义子语言所用的 V_PRV 类型。

```
113c <value types 113c>≡
  #define V_NUM 1
  #define V_STR 2
  #define V_NOD 3
  #define V_PVL 4
  #define V_PRV 5
Defines:
  V NOD, used in chunks 108, 114c, and 116.
  V NUM, used in chunks 108, 114c, and 117-19.
  V PRV, never used.
  V PVL, used in chunks 117a and 118a.
  V STR, used in chunks 108 and 114c.
```

例程 Val 用于表达一个值，同时也是上述类型的并集。

```
114a <value structure 114a>≡
  typedef struct Val {
    int type;
    union { double d;
            char   *s;
            Node   *n;
            void   *v;
    } u;
  } Val;
Defines:
  Val, used in chunks 107c and 112-19.
Uses Node 115b.
```

例程 pv_make 负责构建名称和值确定的一个参数。

```
114b <pv make 114b>≡
  Pval *pv_make(char *name, Val v)
  {
    Pval *pv = (Pval *)malloc(sizeof(Pval));
    pv->name = name;
    pv->val = v;
    pv->next = (Pval *)0;
    return pv;
  }
Defines:
  pv_make, used in chunks 108 and 116b.
Uses Pval 113b and Val 114a.
```

例程 pv_value 返回一个特定类型的值。

```
114c <pv value 114c>≡
  Val pv_value(int type, double num, char *str, Node *nl)
  {
    Val v;
    switch (v.type = type) {
    case V_STR: v.u.s = str; break;
    case V_NUM: v.u.d = num; break;
    case V_NOD: v.u.n = nl; break;
    }
    return v;
  }
Defines:
  pv_value, used in chunks 108 and 119.
Uses Node 115b, V_NOD 113c, V_NUM 113c, V_STR 113c, and Val 114a.
```

例程 pv_append 在参数列表结尾处添加一个参数，用于表达式的操作数列表的构建。

```
115a <pv append 115a>≡
  Pval *pv_append(Pval *pvlist, Pval *pv)
  {
    Pval *p = pvlist;
    if (p == NULL)
      return pv;
    while (p->next ! = NULL)
      p = p->next;
    p->next = pv;
    return pvlist;
  }
Defines:
  pv_append, used in chunks 108 and 116b.
Uses Pval 113b.
```

7.3.5 节点和表达式

表达式对应于一个树形结构，其中，内部节点表示为当前语言的运算符。结构 Node 表示为一个树形节点，并由以下内容构成：指向实现了运算符的函数的指针；定义其操作数的参数列表。

```
115b <node structure 115b>≡
  typedef struct Node {
    struct Val (*func)();
    struct Pval *plist;
  } Node;
Defines:
  Node, used in chunks 107c, 114-16, and 119d.
Uses Pval 113b and Val 114a.
```

例程 t_node 用于构建表达式树的一个节点。

```
115c <node construtor 115c>≡
  Node *t_node(Val (*fun)(), Pval *p)
  {
    Node *n = (Node *) emalloc(sizeof(Node));
    n->func = fun;
    n->plist = p;
    return n;
  }
```

```
Defines:
  t_node, used in chunk 108.
Uses Node 115b, Pval 113b, and Val 114a.
```

当评估一个表达式时，应采用替代模型。具体处理过程将按深度遍历表达式树。为了使当前语言更具通用性，可通过宽度和深度方式遍历树形结构，进而完成表达式的评估过程。通过这种方式，将访问每个树节点，并执行两次实现了运算符的函数。其中，第一次对应于预处理（T_PREP 动作），第二次则对应于自身的执行过程（T_EXEC 动作）。

```
#define T_PREP 0
#define T_EXEC 1
```

而且，我们实现了两种类型的表达式评估，即有损型（destructive）评估和无损型（non-destructive）评估。相应地，例程 t_eval 执行有损型表达式评估，并利用其递归评估过程中的值替换子表达式。

```
116a <eval 116a>≡
  Val t_eval(Node *n)
  {
    Pval *p;

    (*n->func)(T_PREP, n->plist);
    for (p = n->plist; p != NULL; p = p->next)
      if (p->val.type == V_NOD)
        p->val = t_eval(p->val.u.n);
    return (*n->func)(T_EXEC, n->plist);
  }
Defines:
  t_eval, used in chunk 119b.
Uses Node 115b, Pval 113b, T_EXEC, T_PREP, V_NOD 113c, and Val 114a.
```

例程 t_nd_eval 则生成无损型表达式评估，保留子表达式的树形结构，并生成隔离于已评估操作数（针对运算符的最后一个操作数）的列表。

```
116b <non destructive eval 116b>≡
  Val t_nd_eval(Node *n)
  {
    Pval *p, *qlist = NULL;

    (*n->func)(T_PREP, n->plist);
    for (p = n->plist; p != NULL; p = p->next)
      qlist = pv_append(qlist, pv_make(p->name,
```

```
                (p->val.type == V_NOD)? t_nd_eval(p->val.u.n) : p->val));
      return (*n->func)(T_EXEC, qlist);
  }
Defines:
  t_nd eval, used in chunk 119c.
Uses Node 115b, pv_append 115a, pv_make 114b, Pval 113b, T_EXEC, T_PREP,
V_NOD 113c, and Val 114a.
```

例程 **t_pvl** 表示为实现了运算符"列表"的函数,对应于语法结构{ pv, pv, ... }。需要注意的是,这也是实现于扩展语言内核中的唯一语义。

```
117a <pvl 117a>≡
  Val t_pvl(int c, Pval *pvl)
  {
    Val v;
    v.type = V_PVL; v.u.v = pvl;
    return v;
  }
Defines:
  t_pvl, used in chunk 108.
Uses c 109c, Pval 113b, V_PVL 113c, and Val 114a.
```

7.3.6 辅助函数

此处定义了多个辅助函数并与参数列表协同工作。其中,例程 **pvl_to_array** 将数值列表转换为一个数组。

```
117b <pvlto array 117b> ≡
  void pvl_to_array(Pval *pvl, double *a, int n)
  {
    int k = 0;
    Pval *p = pvl;

    while (p != NULL && k < n) {
      a[k++] = (p->val.type == V_NUM) ? p->val.u.d : 0;
      p = p->next;
    }
  }
Defines:
  pvl_to_array, used in chunk 117c.
Uses Pval 113b and V_NUM 113c.
```

例程 pvl_to_v3 将包含 3 个数字的列表转换为一个三维向量。

```
117c  <pvlto v3 117c>≡
  Vector3 pvl_to_v3(Pval *pvl)
  {
    double a[3] = {0, 0, 0};
    pvl_to_array(pvl, a, 3);
    return v3_make(a[0], a[1], a[2] );
  }
Defines:
  pvl_to_v3, used in chunk 118a.
Uses Pval 113b and pvl_to_array 117b.
```

例程 pvl_get_v3 从 pname 标识的列表中析取一个参数。如果参数不存在，该例程将返回默认值 defval。

```
118a  <pvl get v3 118a>≡
  Vector3 pvl_get_v3(Pval *pvl, char *pname, Vector3 defval)
  {
    Pval *p;
    for (p = pvl; p ! = (Pval *)0; p = p->next)
      if (strcmp(p->name, pname) == 0 && p->val.type == V_PVL)
        return pvl_to_v3(p->val.u.v);
    return defval;
  }
Defines:
  pvl_get_v3, never used.
Uses Pval 113b, pvl_to_v3 117c, and V_PVL 113c.
```

例程 pvl_get_num 从列表中析取一个标量值参数。

```
118b  <pvlgetnum 118b>≡
  Real pvl_get_num(Pval *pvl, char *pname, Real defval)
  {
    Pval *p;
    for (p = pvl; p ! = (Pval *)0; p = p->next)
      if (strcmp(p->name, pname) == 0 && p->val.type == V_NUM)
        return p->val.u.d;
    return defval;
  }
Defines:
  pvl_get_num, never used.
Uses Pval 113b and V_NUM 113c.
```

7.4 子语言和应用程序

本节将讨论如何利用扩展语言中的资源定义面向应用程序的子语言。

7.4.1 基于扩展语言的接口

当确定某个子语言时，需要定义实现了期望语义的运算符。对此，例程 lang_defun 用于定义一个新的运算符，并在符号列表中"安装"该运算符的名称，以及实现了该运算符的函数。

```
118c <langdefun 118c>≡
  void lang_defun(char *name, Val (*func)())
  {
    if (sym_lookup(name))
      fprintf(stderr,"lang: symbol %s already defined\n", name);
    else
      sym_install(name, CLASS, func);
  }
Defines:
  lang_defun, never used.
Uses sym_install 112c, sym_lookup 113a, and Val 114a.
```

此处调用词法分析器执行源代码分析，并生成表达式树。

例程 lang_parse 执行 yacc 生成的语法分析器，并依次调用词法分析器。

```
119a <langparse 119a>≡
  int lang_parse()
  {
    return yyparse();
  }
Defines:
  lang_parse, never used.
```

当运行上述程序时，将对表达式树进行评估，其根节点存储于内部变量 root 中。例程 lang_eval 负责运行当前程序并采用有损型方式对树形结构进行评估。

```
119b <langeval 119b>≡
  Val lang_eval()
  {
```

```
    return (root ! = NULL) ? t_eval(root) : pv_value(V_NUM, 0, NULL, NULL);
  }
Defines:
  lang_eval, never used.
Uses pv_value 114c, root, t_eval 116a, V_NUM 113c, and Val 114a.
```

例程 lang_nd_eval 运行当前程序，并采用无损的方式评估树形结构。

```
119c <langndeval 119c>≡
  Val lang_nd_eval(void)
  {
    return (root ! = NULL) ? t_nd_eval(root) : pv_value(V_NUM, 0, NULL, NULL);
  }
Defines:
  lang_nd_eval, never used.
Uses pv_value 114c, root, t_nd_eval 116b, V_NUM 113c, and Val 114a.
```

例程 lang_ptree 返回表达式树。

```
119d <langparse tree 119d>≡
  Node *lang_ptree(void)
  {
    return root;
  }
Defines:
  lang_ptree, never used.
Uses Node 115b and root.
```

7.4.2 实现语义

当实现语言语义时，应通过执行 T_PREP 和 T_EXEC 操作的函数来定义运算符。此类函数包含如下结构：

```
Val f(int call, Pval *pl)
{
  Val v;

  if (call == T_EXEC) {
    /* execute and return value */
  } else if (call == T_PREP) {
    /* preprocess and return null */
  }
```

```
    return v;
}
```

7.4.3 生成解释器

这里需要创建主程序进而生成语言解释器，其中包含了扩展语言内核以及子语言。相应地，程序可划分为以下 3 个步骤：

（1）子语言运算符定义。
（2）源代码编译。
（3）执行程序。

下列程序代码显示了解释器的结构示例。

```
main(int argc, char **argv)
{
 lang_define("f", f);
 ...

 if (lang_parse() == 0)
   lang_eval();

 exit(0);
}
```

7.5 补充材料

本章讨论了与语言和过程描述相关的概念，同时还介绍了扩展语言，并确定三维场景数据。最后，本章开发了一个库，以对该语言提供计算支持。

读者可参考 Aho 和 Ullman 编写的相关著作[Aho and Ullman 79]以了解更多内容。

7.5.1 修正

当前语言库的 API 包含了下列例程：

```
Pval *pv_make(char *name, Val v);
Pval *pv_append(Pval *pvlist, Pval *pv);
Val pv_value(int type, double num, char *str, Node *nl);
```

```
Node *t_node(Val (*fun)(), Pval *p);
Val t_eval(Node *n);
Val t_pvl(int c, Pval *pl);

Node *yyptree();

void lang_define(char *name, Val (*func)());
int lang_parse();
Val lang_eval();

void pvl_to_array(Pval *pvl, double *a, int n);
Vector3 pvl_to_v3(Pval *pvl);
Real pvl_get_num(Pval *pvl, char *pname, Real defval);
Vector3 pvl_get_v3(Pval *pvl, char *pname, Vector3 defval);
```

7.5.2 扩展

本章所讨论的表达式语言功能非常强大，几乎包含了编程语言中的所有组成部分，但唯一的缺憾是无法利用该语言自身创建新的函数。当前语言的函数在 C 语言中定义为基本运算符，同时支持高效的实现。

但是，扩展语言可通过定义宏和函数结构得到进一步的增强。在当前示例中，可使用 C 语言中的预处理器 CPP。通过这种方式，可在当前语言中设置 define 命令。CPP 支持以下格式的宏定义：

```
#define M(P1, P2, PN) macro code
```

预处理程序标识所有已定义的宏并执行宏展开操作，用相应的参数替换文本中的宏调用。例如，宏调用 M (the, b,c)被宏代码所替换，并用 a、b、c 替换 P1、P2、P3。

CPP 的另一个优点是可通过命令 include 包含多个文件。

定义函数的结构需要对语言进行重大改进。这可通过 defun 结构予以实现，defun 将表达式与语言函数的名称（运算符）相关联。

7.5.3 相关信息

Lisp、TCL 和 Moon 中均包含了一些重要的扩展语言示例。作为当前语言的语法基础，其他场景描述语言（除了 VRML 和 OpenInventor）还包括 Geomview（OOGL 和 OFF 格式）、Renderman、PovRay 和 Radiance。

7.6 本章练习

（1）使用 LANG 库针对计算器构建一种语言，同时需要实现以下各项操作：加法、减法、乘法和除法。

（2）扩展（1）中的计算器，以支持命令行中传递的值变量。对此，可采用 arg { var = num } 这一格式。其中，var 表示为变量名，num 表示为默认值，当变量未出现于命令行时将使用该值。

第 8 章 三维几何体模型

三维场景由三维对象集构成。从这一点来看，对象的形式是计算机图形学处理和应用中一项较为重要的内容。本章将讨论三维对象的模型和表达方式。

8.1 建模基础知识

再次强调，本节将使用 4 种环境范例。在物理环境中，需要描述三维对象的形式。对此，需要确定与此类几何模型相关联的模式和参数。在实现过程中，可构建相应的数据结构和过程，进而在计算机设备中支持这一类表达结果，如图 8.1 所示。

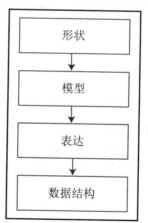

图 8.1 结合体建模中的 4 种抽象级别

8.1.1 模型和几何体描述

对象的几何支撑包括任意空间 $S = \{p \subset \mathbb{R}^3 : p \in O\}$ 内的一组点集。为了更好地刻画点集所描述的几何体，需要设置某些限制条件。

（1）流形。这里，假设对象由齐次点集构成，此类点集可建模为维度 1（曲线）、2（表面）和 3（实体）的流形，并嵌入三维欧几里得空间中。这意味着，在本地中，n 维点集同胚于 $\mathbb{R}^n$ 中的一个开放球体，如图 8.2 所示。

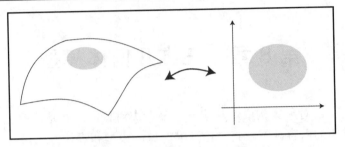

图 8.2 保持既定空间拓扑属性的映射

（2）参数和隐式描述。当对几何体对象选择了数学模型后，需要对其加以具体指定。换而言之，需要通过数学工具描述 n 维流形。对此，可采用相应的函数式描述，也就是说，利用函数指定对象点集。相应地，存在两种几何体函数描述方式，即参数式和隐式。

在参数描述中，点集 $p \in S$ 直接通过函数 $p = f(u)$ 加以定义，该函数利用对象的维度 m 定义于参数空间内，即 $u \in U \subset \mathbb{R}^m$。图 8.3 显示了针对参数曲线的这一概念。

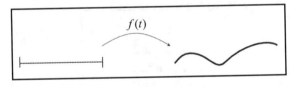

图 8.3 参数曲线

在隐式描述中，几何体间接定义为满足方程 $g(p) = c$ 的点集 $p \in \mathbb{R}^3$。通过这一方式，$S = g^{-1}(c)$ 表示为 g 的逆图像。图 8.4 显示了针对隐式曲线的这一概念。

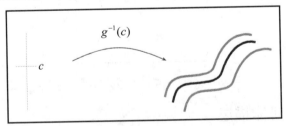

图 8.4 隐式曲线

需要注意的是，参数和隐式描述之间具有互补性。对于 S 中的点，参数描述支持枚举机制；而在与 S 相关的环境空间中，隐式描述则支持分类机制。

另一项较为重要的观察结果是，除了原函数之外（参数和隐式函数），通过某种方式纳入此类函数操作进而生成函数组合，还可进一步提升建模系统的功能。

8.1.2 表达模式

根据之前讨论的函数几何体模型,可通过某种模式并向对象的形式。其中,较为常见的表达模式如下:
- 图元族。
- 构建模式。
- 分解模式。

(1) 图元表达。在图元表达模式中,可定义一组函数表述对象类。除此之外,还可包含相应的几何转换操作(参见第 4 章)。基于图元的建模系统实现意味着针对形状、支持函数和转换操作采用某个库中的定义。

【例 8.1】 (球体族)这种类型的图元系统可用于对大理石进行建模。下面考查此类系统在 4 个环境范例中的构建方式。
- 对象:球体形状。
- 模型:函数式描述。其中,参数形式表示为 $(x, y, z) = r(\cos u \cos v, \sin u \cos v, \sin v)$;隐式形式表示为 $y^2 + z^2 = r^2$。
- 表达方式:描述的参数,即名称和半径(id 和 r)。
- 数据结构:关联表,即 $id_k \rightarrow r_k$, $k = 1, \dots M$。

(2) 构建表达。在构建表达模式中,可使用几何体图元和点集组合操作。据此,可根据简单的图元对象构建更加复杂的组合对象。组合运算符定义了一个代数。

当实现一个构建模型系统时,需要定义图元和组合运算符。系统中的对象表达可通过代数表达式确定。

【例 8.2】 (CSG——构建实体几何)CSG 系统基于点集间并集、交集和差集的布尔操作。根据 4 种环境范例,基于球体的 CSG 系统应涵盖以下内容:
- 对象:球体形状组合。
- 模型:图元为球体;运算符为 ∪、∩、/。
- 表达方式:CSG 布尔表达式。
- 数据结构:表达式的二叉树。

(3) 基于分解的表达方式。在基于分解的表达模式中,可采用与构建模式相反的策略。也就是说,不使用简单的形状构建复杂的形状,而是将复杂形状分解为简单的形状,进而可利用参数或隐式形式加以描述。因此,在该模式中,需要针对各部分组装定义相应的操作。总体而言,可对模型的拓扑元素(顶点、边、表面和壳层)进行分层。组装过程由描述这些元素间的入射关系给出,如图 8.5 所示。

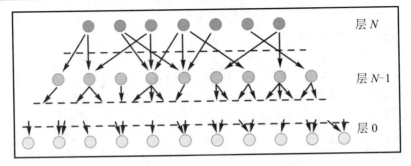

图 8.5 组装模型拓扑元素的入射关系图

基于分解模式的建模系统的实现意味着需要定义所使用的面片（patch）类，以及面片间的组装类型。

【例 8.3】 （多面体）面元（faceted）对象可分解为多边形面片。该表达方式可视为一个多边形网格。在 4 种环境下的范例中，下面考查如何构建此类系统。

- 对象：面元形状。
- 模型：几何的线性分段分解。
- 表达方式：作为多边形网格的表面。
- 数据结构：顶点和面元列表。

8.2 几何图元

作为几何建模系统的基础之一，我们将使用图元族。这里仅对有效对象类提出两项要求，且应具有双重功能描述，即参数化和隐式功能。同时，还应支持预置功能集。通过这种方式，当在系统中包含一个新的图元族时，则有必要实现此类功能。

这一类基本图元包括球体、锥体、圆柱体、二次曲面和超二次曲面、圆环体、盒体和高度曲面。此类对象的表达方式由特定于每种模型族的参数集构成。例如，球体的中心和半径。相关操作包含了相应的几何转换操作，例如空间中的刚体运动。

除此之外，还可将每种图元与包围体关联，并与主方向对齐。该信息通过结构 Box3d 予以定义，表明左下和右上顶点。

```
127a <box3d struct 127a>≡
  typedef struct Box3d {
    Vector3 ll, ur;
  } Box3d;
Defines:
 Box3d, used in chunks 127-29 and 135a.
```

8.2.1 图元对象定义

图元模型的实现应遵循面向对象方案。其中，系统中的图元表达方式采用了由支持函数、包围体、直接与逆转换矩阵和对象参数构成的数据结构 Prim。

```
127b <prim struct 127b>≡
  typedef struct Prim {
    struct PrimFuncs *f;
    Box3d b;
    Matrix4 td, ti;
    void *d;
  } Prim;
Defines:
  Prim, used in chunks 128-40.
Uses Box3d 127a and PrimFuncs 128a.
```

图元的支持函数实现了对象和模型系统间的接口。此类函数对应于图元模型所实现的操作集。通过这一方式，面向对象技术使得每个图元族的具体实现独立于建模系统的其余部分。

结构 PrimFuncs 包含了指向每个图元族函数的指针。

```
128a <prim funcs 128a>≡
  typedef struct PrimFuncs {
    Prim *(*instance)();
    void (*destroy)();
    Box3d (*bbox)();
    int (*classify)();
    Vector3 (*point)();
    Vector3 (*normal)();
    Vector3 (*gradient)();
    Inode *(*intersect)();
    Prim *(*transform)();
    Poly *(*uv_decomp)();
    Vector3 (*texc)();
    Vector3 (*du)();
    Vector3 (*dv)();
    Matrix4 (*local)();
    int (*id)();
    void (*write)();
    void (*draw)();
  } PrimFuncs;
```

```
Defines:
  PrimFuncs, used in chunks 127b, 128b, and 133.
Uses Box3d 127a, Poly 142a, and Prim 127b.
```

8.2.2 泛型接口

面向对象程序设计需要针对全部类成员定义一个泛型接口，并针对每个图元族封装特定的支持函数。此类函数的实现过程较为简单：简单地调用与作为参数传递的图元所对应的函数即可。

接下来，将对每个泛型函数进行描述，指定每个函数完成的操作。例程 prim_instance 和 prim_destroy 分别表示为图元对象的构造函数和析构函数。注意，构造函数将利用默认参数生成一个图元实例。

```
128b <prim instance 128b>≡
  Prim *prim_instance(PrimFuncs *f)
  {
    return f->instance(f);
  }
Defines:
  prim_instance, never used.
Uses Prim 127b and PrimFuncs 128a.

129a <prim destroy 129a>≡
  void prim_destroy(Prim *p)
  {
    (*p->f->destroy)(p);
  }
Defines:
  prim_destroy, never used.
Uses Prim 127b.
```

例程 prim_bbox 返回图元的包围盒。

```
129b <prim bbox 129b>≡
  Box3d prim_bbox(Prim *p)
  {
    return (*p->f->bbox)(p);
  }
Defines:
  prim_bbox, never used.
Uses Box3d 127a and Prim 127b.
```

例程 prim_classify 相对于图元对点 q 进行划分，并使用隐式描述确定该点是否位于对象的内部、外部或边界上。

```
129c <prim classify 129c>≡
  int prim_classify(Prim *p, Vector3 q)
  {
    return (*p->f->classify)(p, q);
  }
Defines:
  prim_classify, never used.
Uses Prim 127b.
```

该例程根据下列代码返回划分结果。

```
#define PRIM_IN   1
#define PRIM_OUT -1
#define PRIM_ON   0
```

例程 prim_point 使用参数描述执行图元点的枚举操作。

```
129d <prim point 129d>≡
  Vector3 prim_point(Prim *p, Real u, Real v)
  {
    return (*p->f->point)(p, u, v);
  }
Defines:
  prim_point, never used.
Uses Prim 127b.
```

例程 prim_normal 计算对应于参数 (u, v) 的表面的法线向量。

```
130a <prim normal 130a>≡
  Vector3 prim_normal(Prim *p, Real u, Real v)
  {
    return (*p->f->normal)(p, u, v);
  }
Defines:
  prim_normal, never used.
Uses Prim 127b.
```

例程 prim_gradient 返回点 q 处、图元隐式函数的梯度。

```
130b <prim gradient 130b>≡
  Vector3 prim_gradient(Prim *p, Vector3 q)
  {
```

```
        return (*p->f->gradient)(p, q);
    }
Defines:
  prim_gradient, never used.
Uses Prim 127b.
```

例程 prim_intersect 利用隐式描述计算射线与图元的交点。

```
130c <prim intersect 130c>≡
  Inode *prim_intersect(Prim *p, Ray r)
    {
        return (*p->f->intersect)(p, r);
    }
Defines:
  prim_intersect, never used.
Uses Prim 127b.
```

例程 prim_transform 针对图元应用几何转换。需要注意的是，矩阵 md 定义了直接转换，而矩阵 mi 则定义了逆转换。

```
130d <prim transform 130d>≡
  Prim *prim_transform(Prim *p, Matrix4 md, Matrix4 mi)
    {
        return (*p->f->transform)(p, md, mi);
    }
Defines:
  prim_transform, never used.
Uses Prim 127b.
```

例程 prim_uv_decomp 生成图元的多边形近似结果，并返回多边形列表（详细内容参见 8.3 节）。

```
131a <prim uvdecomp 131a>≡
  Poly *prim_uv_decomp(Prim *p, Real level)
    {
        return (*p->f->uv_decomp)(p, level);
    }
Defines:
  prim_uv_decomp, never used.
Uses Poly 142a and Prim 127b.
```

例程 prim_texc 计算图元的纹理坐标，并标准化至 [0,1] 区间，对应于 (u, v) 所定义的参数坐标。

第 8 章 三维几何体模型

```
131b <prim texc 131b>≡
  Vector3 prim_texc(Prim *p, Real u, Real v)
  {
    return (*p->f->texc)(p, u, v);
  }
Defines:
  prim_texc, never used.
Uses Prim 127b.
```

例程 prim_du 和 prim_dv 计算与 u 和 v 相关的参数化函数的偏导数。

```
131c <prim du 131c>≡
  Vector3 prim_du(Prim *p, Real u, Real v)
  {
    return (*p->f->du)(p, u, v);
  }
Defines:
  prim_du, never used.
Uses Prim 127b.
```

```
131d <prim dv 131d>≡
  Vector3 prim_dv(Prim *p, Real u, Real v)
  {
    return (*p->f->dv)(p, u, v);
  }
Defines:
  prim_dv, never used.
Uses Prim 127b.
```

例程 prim_local 返回转换矩阵，该矩阵执行图元全局坐标系和局部坐标系间的坐标转换。

```
132a <prim local 132a>≡
  Matrix4 prim_local(Prim *p)
  {
    return (*p->f->local)(p);
  }
Defines:
  prim_local, never used.
Uses Prim 127b.
```

例程 prim_id 返回图元所属族的标识符。

```
132b <prim id 132b>≡
  int prim_id(Prim *p)
  {
```

```
    return p->type;
  }
Defines:
  prim_id, never used.
Uses Prim 127b.
```

例程 prim_write 使用 3D 场景语言描述将图元的表达形式写入 fp 给出的文件中。

```
132c <prim write 132c> ≡
  void prim_write(Prim *p, FILE *fp)
  {
    (*p->f->write)(p, fp);
  }
Defines:
  prim_write, never used.
Uses Prim 127b.
```

除了上述函数之外,每个图元族还定义了一个构建图元的函数。

8.2.3 图元示例

本节将展示如何定义模型系统中的图元对象,并通过球体(位于原点处的球体 S^2)图元考查相关示例。在这一类较为典型的模型中,将采用平移和缩放转换以及任意中心位置和半径得到相应的球体实例。该策略的优点在于,计算过程可在图元的局部坐标系中简单而高效地予以执行。

数据结构 Sphere 封装了球体图元的中心位置和半径。

```
133a <sphere struct 133a>≡
  typedef struct Sphere {
    Vector3 c;
    double r;
  } Sphere;
Defines:
  Sphere, used in chunks 133c, 134b, and 140b.
```

表 sphere_func 定义了球体图元中的各项功能,对应元素包括指向函数的指针,相关函数实现了图元的各种基本操作,稍后将对此加以讨论。

```
133b <sphere funcs 133b>≡
  PrimFuncs sphere_funcs = {
    sphere_instance,
    sphere_destroy,
    sphere_bbox,
```

```
        sphere_classify,
        sphere_point,
        sphere_normal,
        sphere_gradient,
        sphere_intersect,
        sphere_transform,
        sphere_uv_decomp,
        sphere_texc,
        sphere_du,
        sphere_dv,
        sphere_local,
        sphere_id,
        sphere_write,
        sphere_draw,
    };
```
Defines:
  sphere_funcs, used in chunk 140a.
Uses PrimFuncs 128a, sphere_bbox 135a, sphere_classify 135b, sphere_destroy 134a, sphere_draw, sphere_du 139b, sphere_dv 139c, sphere_gradient 136c, sphere_id 139e, sphere_instance 133c, sphere_intersect 137, sphere_local 139d, sphere_normal 136b, sphere_point 136a, sphere_texc 139a, sphere_transform 138a, sphere_uv_decomp 138b, and sphere_write 140b.

例程 sphere_instance 定义为球体图元的构造函数，该函数将分配数据结构，并返回球心 $c = (0, 0, 0)$ 且半径为 $r=1$ 的标准球体。

```
133c <sphere instance 133c>≡
  Prim *sphere_instance(PrimFuncs *f)
  {
    Vector3 ll = {-1,-1,-1}, ur = {1,1,1};
    Prim *p = NEWSTRUCT(Prim);
    Sphere *s = NEWSTRUCT(Sphere);

    p->f = f;
    p->b.ll = ll; p->b.ur = ur;
    p->ti = p->td = m4_ident();
    s->c = v3_make(0,0,0); s->r = 1;
    p->d = s;
    return p;
  }
```
Defines:
  sphere_instance, used in chunks 133b and 140a.
Uses Prim 127b, PrimFuncs 128a, and Sphere 133a.

例程 sphere_destroy 定义为球体图元实例的析构函数，并释放 sphere_instance 所分配的内存空间。

```
134a <sphere destroy 134a> ≡
  void sphere_destroy(Prim *p)
  {
    free(p->d);
    free(p);
  }
Defines:
  sphere_destroy, used in chunk 133b.
Uses Prim 127b.
```

例程 sphere_set 负责修改现有球体的参数。需要注意的是，该例程使用了平移和缩放转换计算图元局部坐标系中坐标的变化。此外，该例程还将更新图元的包围盒。

```
134b <sphere set 134b>≡
  Prim *sphere_set(Prim *p, Vector3 c, double r)
  {
    Sphere *s = p->d;
    s->c = c; s->r = r;
    p->td = m4_m4prod(m4_translate(c.x,c.y,c.z),m4_scale(r,r,r));
    p->ti = m4_m4prod(m4_scale(1/r,1/r,1/r), m4_translate(-c.x,-c.y,-c.z));
    p->b = sphere_bbox(p);
    return p;
  }
Defines:
  sphere_set, used in chunk 140a.
Uses Prim 127b, Sphere 133a, and sphere_bbox 135a.
```

例程 sphere_bbox 针对球体图元生成一个包围盒。首先，该例程将包含标准球体的立方体顶点调整至全局坐标系中，随后计算该立方体的包围盒。

```
135a <sphere bbox 135a>≡
  Box3d sphere_bbox(Prim *p)
  {
    Box3d b;
    Vector3 v;
    double x, y, z;

    for (x = -1; x <= 1; x +=2) {
      for (y = -1; y <= 1; y +=2) {
        for (z = -1; z <= 1; z +=2) {
          v = v3_m4mult(v3_make(x, y, z), p->td);
```

```
              if (x == -1 && y == -1 && z == -1) {
                b.ll = b.ur = v;
              } else {
                if (v.x < b.ll.x) b.ll.x = v.x;
                if (v.y < b.ll.y) b.ll.y = v.y;
                if (v.z < b.ll.z) b.ll.z = v.z;
                if (v.x > b.ur.x) b.ur.x = v.x;
                if (v.y > b.ur.y) b.ur.y = v.y;
                if (v.z > b.ur.z) b.ur.z = v.z;
              }
            }
          }
        }
        return b;
      }
Defines:
  sphere_bbox, used in chunks 133b and 134b.
Uses Box3d 127a and Prim 127b.
```

例程 sphere_classify 执行点-集合分类计算。首先将点 q 转换至球体的局部坐标系中，随后采用隐式方程 $x^2 + y^2 + z^2 - 1 = 0$ 判断点 $q = (x, y, z)$ 是否位于球体的内部、外部或表面上。

```
135b <sphere classify 135b>≡
    int sphere_classify(Prim *p, Vector3 q)
    {
      Vector3 w = v3_m4mult(q, p->ti);
      Real d = v3_sqrnorm(w);
      return (d < 1)? PRIM_IN : ((d > 1)? PRIM_OUT : PRIM_ON);
    }
Defines:
  sphere_classify, used in chunk 133b.
Uses Prim 127b, PRIM_IN, PRIM_ON, and PRIM_OUT.
```

例程 sphere_point 返回与参数坐标(u, v)所对应的、球体上的一点。其中，所选参数使用了球体坐标，且有 $u \in [0, 2\pi]$，$v \in [-\pi/2, \pi/2]$。需要注意的是，该参数化行为在球体的两极（$v = \pm\pi/2$）处是奇异的。此外，计算过程在局部坐标系中执行，而最终结果将转换至全局坐标系中。

```
136a <sphere point 136a> ≡
  Vector3 sphere_point(Prim *p, Real u, Real v)
  {
    Vector3 w;
```

```
    w.x = cos(u)*cos(v);
    w.y = sin(u)*cos(v);
    w.z = sin(v);
    return v3_m4mult(w, p->td);
  }
Defines:
  sphere_point, used in chunk 133b.
Uses Prim 127b.
```

例程 sphere_normal 计算点 $w = f(u, v)$ 处垂直于球体的向量。该向量位于局部坐标系中，且需要转换至全局坐标系中（参见第 4 章）。

```
136b <sphere normal 136b> ≡
  Vector3 sphere_normal(Prim *p, Real u, Real v)
  {
    Vector3 w;
    w.x = cos(u)*cos(v);
    w.y = sin(u)*cos(v);
    w.z = sin(v);
    return v3_m4mult(w, m4_transpose(p->ti));
  }
Defines:
  sphere_normal, used in chunk 133b.
Uses Prim 127b.
```

例程 sphere_gradient 返回与球体 $\nabla f = \left(\dfrac{f}{x}, \dfrac{f}{y}, \dfrac{f}{z} \right) = (2x, 2y, 2z)$ 关联的、隐式函数 $f(x, y, z) = x^2 + y^2 + z^2 - 1$ 点 q 处的梯度。在计算之前，点将被转换至球体的局部坐标系中。

```
136c <sphere gradient 136c>≡
  Vector3 sphere_gradient(Prim *p, Vector3 q)
  {
    Vector3 w = v3_scale(2.0, v3_m3mult(q, p->ti));
    return v3_m3mult(w, m4_transpose(p->ti));
  }
Defines:
  sphere_gradient, used in chunks 133b and 137.
Uses Prim 127b.
```

例程 sphere_intersect 计算射线 s 与球体间的交点。首先，射线被转换至局部坐标系中，随后将射线 $r(t) = o + td$ 的参数方程代入球体的隐式方程 $x^2 + y^2 + z^2 - 1 = 0$ 中，进而得到 t 的二次方程。

$$at^2 + 2bt + c - 1 = 0 \qquad (8.1)$$

其中，$a = d_x^2 + d_y^2 + d_z^2$，$b = o_x d_x + o_y d_y + o_z d_z$，$c = o_x^2 + o_y^2 + o_z^2$。

通过计算上述方程的根即可得到射线与球体间的交点，即满足式（8.1）的、$r(t)$ 的参数 t_0 和 t_1。

```
137 <sphere intersect 137> ≡
  Inode *sphere_intersect(Prim *p, Ray rs)
  {
    double a, b, c, disc, t0, t1;
    Inode *in, *out;
    Ray r = ray_transform(rs, p->ti);

    a = SQR(r.d.x) + SQR(r.d.y) + SQR(r.d.z);
    b = 2.0 * (r.d.x * r.o.x + r.d.y * r.o.y + r.d.z * r.o.z);
    c = SQR(r.o.x) + SQR(r.o.y) + SQR(r.o.z) - 1;
    if ((disc = SQR(b) - 4 * a * c) <= 0)
      return (Inode *)0;
    t0 = (-b + sqrt(disc)) / (2 * a);
    t1 = (-b - sqrt(disc)) / (2 * a);
    if (t1 < RAY_EPS)
      return (Inode *)0;
    if (t0 < RAY_EPS) {
      Vector3 n1 = v3_unit(sphere_gradient(p, ray_point(rs, t1)));
      return inode_alloc(t1, n1, FALSE);
    } else {
      Vector3 n0 = v3_unit(sphere_gradient(p, ray_point(rs, t0)));
      Vector3 n1 = v3_unit(sphere_gradient(p, ray_point(rs, t1)));
      i0 = inode_alloc(t0, n0, TRUE);
      i1 = inode_alloc(t1, n1, FALSE);
      i0->next = i1;
      return i0;
    }
  }
Defines:
  sphere_intersect, used in chunk 133b.
Uses Prim 127b and sphere_gradient 136c.
```

例程 sphere_transform 向球体应用几何转换。注意，这里需要提供直接转换和逆转换矩阵。

```
138a <sphere transform 138a>≡
  Prim *sphere_transform(Prim *p, Matrix4 md, Matrix4 mi)
```

```
    {
      p->td = m4_m4prod(md, p->td);
      p->ti = m4_m4prod(p->ti, mi);
      return p;
    }
Defines:
  sphere_transform, used in chunk 133b.
Uses Prim 127b.
```

通过对参数域进行正则分解，例程 sphere_uv_decomp 生成球体的多边形近似结果。

```
138b <sphere uv decomp 138b>≡
  Poly *sphere_uv_decomp(Prim *p)
  {
    Real u, v, nu = 20, nv = 10;
    Real iu = ULEN/nu, iv = VLEN/nv;
    Poly *l = NULL;

    for (u = UMIN; u < UMAX; u += iu) {
      for (v = VMIN; v < VMAX; v += iv) {
        l = poly_insert(l,
            poly3_make(v3_make(u,v,1),v3_make(u,v+iv,1),v3_make(u+iu,v,1)));
        l = poly_insert(l,
            poly3_make(v3_make(u+iu,v+iv,1),v3_make(u+iu,v,1),
                       v3_make(u,v+iv,1)));
      }
    }
    return l;
  }
Defines:
  sphere_uv_decomp, used in chunk 133b.
Uses Poly 142a, poly3_make 144b, poly_insert 143d, Prim 127b, ULEN, UMAX,
UMIN, VLEN, VMAX, and VMIN.


#define UMIN (0)
#define UMAX (PITIMES2)
#define ULEN (UMAX - UMIN)
#define VEPS (0.01)
#define VMIN ((-PI/2.0) + VEPS )
#define VMAX ((PI/2.0) - VEPS)
#define VLEN (VMAX - VMIN)
```

第 8 章 三维几何体模型

例程 sphere_texc 返回球体的纹理坐标,并标准化至[0, 1]中。

```
139a  <sphere texc 139a>≡
  Vector3 sphere_texc(Prim *p, Real u, Real v)
  {
    return v3_make((u - UMIN)/ULEN, (v - VMIN)/VLEN, 0);
  }
Defines:
  sphere_texc, used in chunk 133b.
Uses Prim 127b, ULEN, UMIN, VLEN, and VMIN.
```

例程 sphere_du 和 sphere_dv 返回球体参数函数的偏导数。

```
139b  <sphere du 139b>≡
  Vector3 sphere_du(Prim *p, Real u, Real v)
  {
    return v3_make(- sin(u) * cos(v), cos(u) * cos(v), 0);
  }
Defines:
  sphere_du, used in chunk 133b.
Uses Prim 127b.
```

```
139c  <sphere dv 139c>≡
  Vector3 sphere_dv(Prim *p, Real u, Real v)
  {
    return v3_make(- cos(u) * sin(v), - sin(u) * sin(v), cos(v));
  }
Defines:
  sphere_dv, used in chunk 133b.
Uses Prim 127b.
```

例程 sphere_local 将转换矩阵返回至球体的局部坐标系中。

```
139d  <sphere local 139d>≡
  Matrix4 sphere_local(Prim *p)
  {
    return p->ti;
  }
Defines:
  sphere_local, used in chunk 133b.
Uses Prim 127b.
```

例程 sphere_id 返回球体图元的类标识符。

```
139e <sphere id 139e>≡
  int sphere_id(Prim *p)
  {
    return SPHERE;
  }
Defines:
  sphere_id, used in chunk 133b.
Uses Prim 127b and SPHERE.
```

三维场景描述语言的球体表达包含下列语法形式：

```
primobj = sphere { center = {1,2,3}, radius = 4 }
```

例程 sphere_parse 执行球体图元的解析操作。

```
140a <sphere parse 140a>≡
  Val sphere_parse(int pass, Pval *pl)
  {
    Val v;

    if (pass == T_POST) {
      Vector3 c = pvl_get_v3(pl, "center", v3_make(0,0,0));
      double r = pvl_get_num(pl, "radius", 1);
      v.type = PRIM;
      sphere_set(v.u.v = sphere_instance(&sphere_funcs), c, r);
    }
    return v;
  }
Defines:
  sphere_parse, never used.
Uses sphere_funcs 133b, sphere_instance 133c, and sphere_set 134b.
```

例程 sphere_write 将球体的表达结果写入某个文件中。

```
140b <sphere write 140b>≡
  void sphere_write(Prim *p, FILE *fp)
  {
    Sphere *s = p->d;
    fprintf(fp, "sphere { \n");
    fprintf(fp, "\t\t center = {%g, %g, %g},\n",s->c.x,s->c.y,s->c.z);
    fprintf(fp, "\t\t radius = %g \n}\n",s->r);
  }
Defines:
  sphere_write, used in chunk 133b.
Uses Prim 127b and Sphere 133a.
```

8.3 表面和多边形网格的近似计算

本节将讨论表面的近似计算问题，该问题与 8.1.2 节中的基于分解的模式表达密切相关。虽然分解模式也可用于准确地表达表面，但表面的近似计算也在分解的基础上进行。相应地，近似表面存在多种解决方案，大多数方案均采用了 n 级多边形面片，两样条则是一种重要的表达方式。

8.3.1 近似方法

表面近似方法源自问题的初始描述，即参数或隐式形式，甚至是通过传感器或模拟得到的密集点采样结果。

根据初始描述，近似方法计算表面的分解结果，其中，每个面片近似于一块表面。

其中，近似方法采用了两种基本操作，即采样和结构化操作。相应地，采样获取表面点并生成面片；结构化操作则涉及面片的拼接以形成网格。取决于采样模式，该过程涵盖了两种方法——均匀法和自适应法。对于结构化操作，则分别包含了以下网格类型：泛型、矩形和单形。

8.3.2 分段式线性近似方法

分段式线性近似是表面近似表达中较为常见的模式。分段近似使得该方法生成分解模型，其优点主要体现在：算法和表达的简单性、软件和硬件的计算支持，以及系统间的兼容性。

表面的分段式线性近似表达方式是多边形网格。多边形网格通过各种拓扑元素间的关联关系体现其拓扑结构，包括顶点、边、面和壳体。几何体则通过顶点的位置予以确定。通过拓扑结构和顶点间的线性插值，整体表面将被重构。

网格的数据结构可通过显式表达的拓扑关系类型进行分类。其中，较为常见的结构包含多边形列表、顶点和面元列表以及边图。

8.4 多边形表面

本节将在建模系统中采用多边形网格作为分解模式。该表达方式支持多面体的真实描述，以及通用表面的近似描述。相应地，所选取的数据结构由顶点坐标数组构成。除了对应位置之外，其他与顶点关联的有用信息还包括表面的法线向量、纹理坐标、颜色和材质（此类信息当前均未予显示）。第 9 章还将针对此类信息表达提供不同的解决方案。

8.4.1 n 边多边形

多边形网格的基本数据结构定义了一个 n 边多边形，同时也是构成网格的单链表元素。

```
142a <poly struct 142a>≡
  typedef struct Poly {
    struct Poly *next;
    int n;
    Vector3 *v;
  } Poly;
Defines:
  Poly, used in chunks 128a, 131a, 138b, and 142–49.
```

例程 poly_alloc 定义为多边形的构造函数。

```
142b <poly alloc 142b>≡
  Poly *poly_alloc(int n)
  {
    Poly *p = NEWSTRUCT(Poly);
    p->n = n;
    p->v = NEWARRAY(n, Vector3);
    p->next = NULL;
    return p;
  }
Defines:
  poly_alloc, used in chunks 144b, 145b, and 149b.
Uses Poly 142a.
```

例程 poly_transform 和 poly_homoxform 分别向多边形顶点应用了仿射转换和投影转换。

```
142c <poly transform 142c>≡
  Poly *poly_transform(Poly *p, Matrix4 m)
  {
    int i;
    for (i = 0; i < p->n; i++)
      p->v[i] = v3_m4mult(p->v[i], m);
    return p;
  }
Defines:
```

第 8 章 三维几何体模型

```
  poly_transform, never used.
Uses Poly 142a.
```

143a <poly homoxform 143a>≡
```
  Poly *poly_homoxform(Poly *p, Matrix4 m)
  {
    int i;
    for (i = 0; i < p->n; i++)
      p->v[i] = v3_v4conv(v4_m4mult(v4_v3conv(p->v[i] ), m));
    return p;
  }
Defines:
  poly_homoxform, never used.
Uses Poly 142a.
```

例程 poly_normal 计算多边形的法线向量。

143b <poly normal 143b>≡
```
  Vector3 poly_normal(Poly *p)
  {
    return v3_unit(v3_cross(v3_sub(p->v[1], p->v[0]), v3_sub(p->v[2], p->v[0])));
  }
Defines:
  poly_normal, used in chunk 146b.
Uses Poly 142a.
```

例程 poly_centr 计算多边形的形心。

143c <poly centr 143c>≡
```
  Vector3 poly_centr(Poly *p)
  {
    int i; Vector3 c = v3_make(0,0,0);
    for (i = 0; i < p->n; i++)
      c = v3_add(c, p->v[i] );
    return v3_scale((Real)(p->n), c);
  }
Defines:
  poly_centr, never used.
Uses Poly 142a.
```

例程 poly_insert 向多边形列表中添加一个多边形。

143d <poly insert 143d> ≡
```
  Poly *poly_insert(Poly *pl, Poly *p)
```

```
    {
      p->next = p1;
      return p;
    }
Defines:
  poly_insert, used in chunks 138b, 148b, and 149b.
Uses Poly 142a.
```

例程 poly_copy 将多边形顶点内容复制至另一个多边形中。

```
144a <poly copy 144a>≡
  int poly_copy(Poly *s, Poly *d)
    {
      int i;
      for (i = 0; i < s->n; i++)
        d->v[i] = s->v[i] ;
      return (d->n = s->n);
    }
Defines:
  poly_copy, never used.
Uses Poly 142a.
```

8.4.2 三角形

三角形是 n 边多边形的特例,其中 $n = 3$。此处将采用结构 Poly 表示三角形,以及专有的例程与其协同工作。

```
144b <poly3 make 144b>≡
  Poly *poly3_make(Vector3 v0, Vector3 v1, Vector3 v2)
    {
      Poly *p = poly_alloc(3);
      p->v[0] = v0; p->v[1] = v1; p->v[2] = v2;
      return p;
    }
Defines:
  poly3_make, used in chunk 138b.
Uses Poly 142a and poly_alloc 142b.
```

这里将使用列表的列表表示三维场景描述语言中的三角形,通过这种方式,三角形可通过 3 顶点(基于(x, y, z)坐标)列表加以确定。具体描述具有以下形式:

```
{ { NUM, NUM, NUM}, { NUM, NUM, NUM}, { NUM, NUM, NUM} }
```

第 8 章 三维几何体模型

例程 poly3_read 从 fp 指定的文件中读取一个三角形。

```
144c  <poly3 read 144c>≡
  int poly3_read(Poly *p, FILE* fp)
  {
    char *fmt = "{%lf, %lf, %lf},";
    int i, n;

    fscanf(fp,"{");
    for (i = 0; i < 3; i++) {
      if ((n=fscanf(fp, fmt,&(p->v[i].x),&(p->v[i].y),&(p->v[i].z))) == EOF)
        return EOF;
      else if (n ! = 3)
        fprintf(stderr,"Error reading polyfile");
      fscanf(fp,"}\n");
    }
    return (p->n = 3);
  }
Defines:
  poly3_read, never used.
Uses Poly 142a.
```

例程 poly3_write 将三角形写入一个文件中。

```
145a  <poly3 write 145a>≡
  void poly3_write(Poly *p, FILE* fp)
  {
    if ( (v3_norm(v3_sub(p->v[0], p->v[1] )) < EPS)
      || (v3_norm(v3_sub(p->v[1], p->v[2] )) < EPS)
      || (v3_norm(v3_sub(p->v[2], p->v[0] )) < EPS))
      fprintf(stderr, "(poly3_write) WARNING: degenerate polygon\n");
    fprintf(fp, "{{%g, %g, %g}, ",  p->v[0].x, p->v[0].y, p->v[0].z);
    fprintf(fp, " {%g, %g, %g}, ",  p->v[1].x, p->v[1].y, p->v[1].z);
    fprintf(fp, " {%g, %g, %g}}\n", p->v[2].x, p->v[2].y, p->v[2].z);
  }
Defines:
  poly3_write, used in chunk 148a.
Uses Poly 142a.
```

例程 poly3_parse 根据场景描述语言中的表达式执行三角形解析操作。

```
145b  <poly3 parse 145b>≡
  Poly *poly3_parse(Pval *plist)
  {
```

```
    Pval *pl;
    int k;

    for (pl = plist, k = 0; pl ! =NULL; pl = pl->next, k++)
       ;
    if (k ! = 3) {
      fprintf(stderr, "(poly3): wrong number of vertices %d\n", k);
      return NULL;
    } else {
      Poly *t = poly_alloc(3);
      for (pl = plist, k = 0; pl ! =NULL; pl = pl->next, k++)
        if (pl->val.type == V_PVL)
          t->v[k] = pvl_to_v3(pl->val.u.v);
        else
          fprintf(stderr, "(poly3): error in vertex\n");
      return t;
    }
  }
Defines:
  poly3_parse, used in chunk 148b.
Uses Poly 142a and poly_alloc 142b.
```

例程 **poly3_area** 计算三角形的面积。

```
146a <poly3 area 146a>≡
  Real poly3_area(Poly *p)
  {
    return v3_norm(v3_cross(v3_sub(p->v[1], p->v[0]), v3_sub(p->v[2], p->v[0])))/2;
  }
Defines:
  poly3_area, never used.
Uses Poly 142a.
```

例程 **poly3_plane** 计算三角形支撑平面方程。

```
146b <poly3 plane 146b>≡
  Vector4 poly3_plane(Poly *p)
  {
    Vector3 n = poly_normal(p);
    Real d = v3_dot(n, p->v[0] );
    return v4_make(n.x, n.y, n.z, d);
  }
Defines:
```

poly3_plane, used in chunk 147a.
Uses Poly 142a and poly_normal 143b.

例程 plane_ray_inter 计算射线和平面间的交点。

```
146c  <poly intersect 146c>≡
  Real plane_ray_inter(Vector4 h, Ray r)
  {
    Vector3 n = {h.x, h.y, h.z};
    Real denom = v3_dot(n, r.d);
    if (REL_EQ(denom, 0))
      return MINUS_INFTY;
    else
      return (h.w + v3_dot(n, r.o)) / denom;
  }
```
Defines:
  plane_ray_inter, used in chunk 147a.

在前述计算的基础上，例程 poly3_ray_inter 计算射线和三角形之间的交点。

```
147a  <poly3 ray inter 147a>≡
  Real poly3_ray_inter(Poly *p, Ray r)
  {
    Vector4 h; Vector3 q0, q1, q2;
    Real t, d, a, b;

    t = plane_ray_inter((h = poly3_plane(p)), r);
    if (t < 0)
      return MINUS_INFTY;

    q0 = v3_sub(ray_point(r, t), p->v[0] );
    q1 = v3_sub(p->v[1], p->v[0] );
    q2 = v3_sub(p->v[2], p->v[0] );

    switch (max3_index(fabs(h.x), fabs(h.y), fabs(h.z))) {
    case 1:
      PROJ_BASE(a, b, q0.y, q0., q1.y, q1.z, q2.y, q2.z); break;
    case 2:
      PROJ_BASE(a, b, q0.x, q0.z, q1.x, q1.z, q2.x, q2.z); break;
    case 3:
      PROJ_BASE(a, b, q0.x, q0.y, q1.x, q1.y, q2.x, q2.y); break;
    }
    if ((a >= 0 && b >= 0 && (a+b) <= 1))
      return t;
```

```
      else
        return MINUS_INFTY;
    }
Defines:
  poly3_ray_inter, never used.
Uses plane_ray_inter 146c, Poly 142a, poly3_plane 146b, and PROJ_BASE 147b.
```

宏 PROJ_BASE 将某个向量投影至两个向量构成的平面上。

```
147b <projbase 147b>≡
  #define PROJ_BASE(A, B, Q0_S, Q0_T, Q1_S, Q1_T, Q2_S, Q2_T) \
    { Real d = (Q1_S * Q2_T - Q2_S * Q1_T); \
      A = (Q0_S * Q2_T - Q2_S * Q0_T) / d; \
      B = (Q1_S * Q0_T - Q0_S * Q1_T) / d; \
    }
Defines:
  PROJ_BASE, used in chunk 147a.
```

8.4.3 三角形列表

在三维场景描述语言中，三角形网格的表达方式遵循以下格式：

```
polyprim = trilist {
          {{3,2,3}, {4,5,6}, {7,8,9}},
          {{5,2,3}, {4,9,6}, {7,2,9}}
          {{1,4,3}, {3,5,6}, {7,3,9}}
          {{7,2,3}, {4,2,6}, {1,8,9}}
          {{6,2,3}, {4,1,6}, {3,8,9}}
          {{8,2,3}, {4,2,6}, {3,8,9}}
          {{1,5,3}, {4,5,3}, {1,7,9}}
}
```

例程 trilist_write 将三角形网格写入某个文件中。

```
148a <trilist_write 148a>≡
  void trilist_write(Poly *tlist, FILE* fp)
  {
    Poly *p = tlist;

    fprintf(fp, "trilist {\n");
    while (p != NULL) {
      poly3_write(p, fp);
      if ((p = p->next) != NULL)
```

第 8 章 三维几何体模型

```
      fprintf(fp, ",\n");
    }
    fprintf(fp, "}\n");
  }
Defines:
  trilist_write, never used.
Uses Poly 142a and poly3_write 145a.
```

例程 trilist_parse 根据三维场景描述语言中的三角形网格解释表达式。

```
148b <trilist parse 148b>≡
  Val trilist_parse(int pass, Pval *plist)
  {
    Val v;

    if (pass == T_POST) {
      Pval *pl;
      Poly *tl = NULL;

      for (pl = plist; pl != NULL; pl = pl->next) {
        if (pl->val.type == V_PVL)
          tl = poly_insert(tl, poly3_parse(pl->val.u.v));
        else
          fprintf(stderr, "(trilist): syntax error\n");
      }
      v.type = POLYLIST;
      v.u.v = tl;
    }
    return v;
  }
Defines:
  trilist_parse, never used.
Uses Poly 142a, poly3_parse 145b, and poly_insert 143d.
```

对于三角形网格，其他辅助例程还包括：计算三角形列表中元素数量的 plist_lenght 例程；针对多边形列表分配内存空间的 plist_alloc 例程。

```
149a <plist lenght 149a>≡
  int plist_lenght(Poly *p)
  {
    int n = 0;
    while (p != NULL) {
      n++; p = p->next;
    }
```

```
    return n;
  }
Defines:
 plist_lenght, never used.
Uses Poly 142a.
```

149b <plist alloc 149b>≡
```
  Poly *plist_alloc(int n, int m)
  {
    Poly *l = NULL;
    while (n--)
      l = poly_insert(l, poly_alloc(m));
    return l;
  }
Defines:
 plist_alloc, never used.
Uses Poly 142a, poly_alloc 142b, and poly_insert 143d.
```

8.5 补充材料

本章讨论了三维几何形状的表达方式，并针对图元和多边形网格构建展示了相关库。图 8.6 显示了几何图元的正交投影绘制示例；图 8.7 显示了多边形网格的正交投影绘制示例。

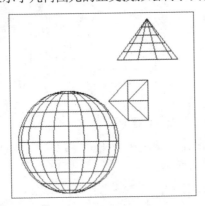

图 8.6　几何图元的正交投影

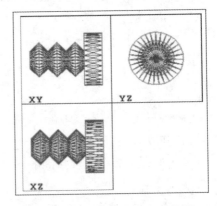

图 8.7　多边形网格的正交投影

图元库 API 包含了下列例程：

```
Prim *prim_instance(int class);
Box3d prim_bbox(Prim *p);
```

```
int prim_classify(Prim *p, Vector3 q);
Vector3 prim_gradient(Prim *p, Vector3 q);
Vector3 prim_point(Prim *p, Real u, Real v);
Vector3 prim_normal(Prim *p, Real u, Real v);
Inode *prim_intersect(Prim *p, Ray r);
Prim *prim_transform(Prim *p, Matrix4 md, Matrix4 mi);
Poly *prim_uv_decomp(Prim *p, Real level);
Vector3 prim_texc(Prim *p, Real u, Real v);
Vector3 prim_du(Prim *p, Real u, Real v);
Vector3 prim_dv(Prim *p, Real u, Real v);
Matrix4 prim_local(Prim *p);
int prim_id(Prim *p);
void prim_write(Prim *p, FILE *fp);
```

多边形网格库 API 包含了下列例程：

```
Poly *poly_alloc(int n);
Poly *poly_transform(Poly *p, Matrix4 m);
Vector3 poly_normal(Poly *p);
Poly *poly_insert(Poly *pl, Poly *p);
Inode *poly_intersect(Poly *p, Vector4 plane, Ray r);
void trilist_write(Poly *tlist, FILE* fp);
Val trilist_parse(int pass, Pval *plist);
```

8.6 本章练习

（1）编写程序，利用场景描述语言扫描和写入图元 SPHERE。
（2）尝试实现新的几何图元，如锥体。
（3）编写程序绘制图元的正交投影。
（4）编写程序测试某个点是否位于实体图元内部。
（5）编写程序计算实体图元的近似体积。提示：可采用光线跟踪技术。
（6）编写程序扫描并写入多边形网格。
（7）编写程序绘制多边形网格的正交投影。
（8）尝试利用顶点和多边形列表结构实现多边形网格描述。
（9）编写程序创建、编辑和转换几何图元。
（10）编写程序创建、编辑和转换多边形网格。

第 9 章 建 模 技 术

第 8 章介绍了数学模型,以描述三维对象几何体以及表达模式,进而在计算机设备上对其予以显示。在几何体建模中,另一个较为重要的问题是,一旦选择了相应的表达模式,应如何确定对象的形式?本章将讨论建模技术,进而对模型的自由度进行更加直观的操控。

9.1 建模系统的基础知识

本节将使用 4 种环境范例阐述建模系统的概念。其中,建模处理始于基于用户界面的物理环境。用户的动作对应于定义于数学环境的方法和几何操作。此类操作将转换为表达环境中的处理过程和建模技术。最后,相关技术映射至实现环境中的建模系统架构中,如图 9.1 所示。稍后将对每个级别加以讨论。

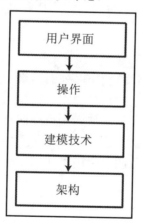

图 9.1 建模系统的抽象级别

9.1.1 用户界面

建模系统中的用户界面与系统指定模型参数的方式相关。对此,存在两种基本的建模模型,即纹理(非交互式)和图形(交互式)。

在非交互式纹理模式中，用户通过命令向系统确定模型。总体而言，这种界面类型基于定义命令格式的语言（语法）及其功能（语义）。对象可通过建模语言中的表达式加以指定。

在交互式图形模式中，用户通过图形界面确定模型，通常实现于包含窗口系统的交互式图形工作站中。该系统包含了模型参数的查看和控制机制。对象通过直接操控建模。

除此之外，还存在上述两种基本界面模式构成的混合系统，如交互式纹理系统，以及基于过程式表达的图形系统。

9.1.2 模型操作

在建模系统中，几何模型基于第 8 章所介绍的表达模式之一，其中涉及图元族、构建模式以及分解模式。

操作集与每种表达相关联，进而生成相关模型。另外，该集合也是模型描述的一个组成部分。利用此类操作，可在建模系统中构建模型，并对现有的对象形式进行调整。其他操作还包括分析对象属性和创建模拟环境。

相应地，建模操作分为两类，即几何操作和组合操作。其中，几何操作涵盖了仿射转换和一般的变形操作；组合操作则包含点集操作和混合操作。

9.1.3 建模技术

如前所述，几何模型可视为建模操作序列的结果。建模技术涉及基于计算过程的表达行为，以及构建特定几何体类型的建模操作。模型组件表示为几何元素，以及针对此类元素的操作。建模技术定义了用于几何元素操作应用的合成规则。最终结果表示为基于表达模式的、有效表达式的对象描述。

需要注意的是，可与同一对象的多个模型协同工作，同时提供了对象几何体的准确或近似表达结果。

9.1.4 系统架构

建模系统架构很大一部分取决于表达模式和所关联的建模技术的选择方案。

对于建模系统来说，存在两种架构选择方案，即基于单一表达的系统，以及基于多重表达的系统。在第二种方案中，取决于表达间的转换，可设置相应的主表达和次级表达，如图 9.2 所示。

第 9 章 建模技术

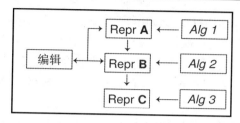

图 9.2　包含多重表达的建模系统

9.2　构造模型

本节将讨论 CSG（构建实体几何体）模型的创建技术。

构建模型基于几何元素以及点集操作。当定义这一类操作时，可使用逻辑运算符以及点成员分类函数，即 inside(C, p)，该函数可确定点 $p \in \mathbb{R}^3$ 是否位于实体 C 的内部、外部或边界处。

- 并集：$A \cup B$ = inside(A) || inside(B)。
- 交集：$A \cap B$ = inside(A) && inside(B)。
- 补集：$\overline{A}$ = inside($\overline{A}$)。
- 差集：$A \backslash B = A \cap \overline{B}$ = inside(A) && inside($\overline{B}$)。

注意，还可使用表实现 CSG 操作。实际上，鉴于差集操作可根据交集实现，因而可针对并集和交集操作定义该表，如表 9.1 和 9.2 所示。

表 9.1　并集操作

∪	in	on	out
in	in	in	in
on	in	?	on
out	in	on	out

表 9.2　交集操作

∩	in	on	out
in	in	on	out
on	on	?	out
out	out	out	out

一种值得关注的情况是，实体边界处点分类的时机往往难以把控。针对这一问题，

需要定义规范化的 CSG 操作[Requicha 80]。

通过前述定义的布尔点成员操作，CSG 对象的模型可视为几何图元的合成结果。其中，对象的外部表达通过 CSG 表达式定义；而对象的内部表达则通过表达式树完成。

注意，前述内容仅定义了 CSG 运算符的语义。当实现该表达时，需要定义对应的语法。

【例 9.1】（CSG 对象）利用中缀语法，CSG 对象的表达式如下所示。

$$A \cup (B - (C \cap D))$$

图 9.3（a）中显示了对象的 CSG 树；图 9.3（b）则显示了最终的几何体结果。

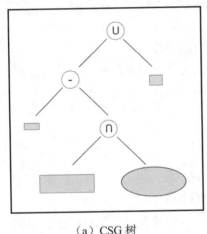
（a）CSG 树

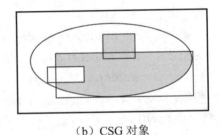
（b）CSG 对象

图 9.3　CSG 树和 CSG 对象

最终，通过观察可知，CSG 建模可视为表达模式和建模技术的组合结果。换而言之，CSG 建模同时体现了构建表达和基于语言的建模技术。

接下来将讨论系统中的 CSG 表达和 CSG 建模技术。

9.2.1　CSG 结构

CSG 模型的内部表达可表示为一棵二叉树。其中，叶节点表示为图元对象，中间节点则表示为 CSG 运算符。CsgNode 结构表示 CSG 树的一个节点。

```
157a <csgnode 157a>≡
  typedef struct CsgNode {      /* CSG Tree Node */
    int type;                   /* CSG_PRIM or CSG_COMP */
    union {
      struct CsgComp c;         /* Composite */
```

第 9 章 建模技术

```
    struct Prim *p;         /* Primitive */
  } u;
} CsgNode;
```
Defines:
  CsgNode, used in chunks 157-60.
Uses CSG_COMP, CSG_PRIM, and CsgComp 157b.

节点可以是类型基本对象（CSG_PRIM）或组合操作（CSG_COMP）。

```
#define CSG_PRIM 0
#define CSG_COMP 1
```

图元对象可通过结构 Prim 定义（参见第 8 章）。CSG 组合操作则通过结构 CsgComp 予以表示，其中包含了操作代码和 CSG 子树给出的操作数。

157b <csg comp 157b>≡
```
  typedef struct CsgComp{         /* CSG Composite */
    char op;                      /* Boolean Operation + - * */
    struct CsgNode *lft, *rgt;    /* Pointer to Children */
  } CsgComp;
```
Defines:
  CsgComp, used in chunk 157a.
Uses CsgNode 157a.

当生成 CSG 树时，将使用两个不同树节点类型的构造操作。另外，例程 csg_prim 将图元对象封装为 CSG 树的叶节点。

157c <csg prim 157c>≡
```
  CsgNode *csg_prim(Prim *p)
  {
    CsgNode *n = (CsgNode *) NEWSTRUCT(CsgNode);
    n->type = CSG_PRIM;
    n->u.p = p;
    return n;
  }
```
Defines:
  csg_prim, used in chunks 159-61.
Uses CSG_PRIM and CsgNode 157a.

例程 csg_link 构建一个组合 CSG 对象，该对象由应用于两棵 CSG 子树上的 CSG 运算符形成。

158a <csg link 158a>≡
```
  CsgNode *csg_link(int op, CsgNode *lft, CsgNode *rgt)
```

```
    {
      CsgNode *n = NEWSTRUCT(CsgNode);
      n->type = CSG_COMP;
      n->u.c.op = op;
      n->u.c.lft = lft;
      n->u.c.rgt = rgt;
      return n;
    }
Defines:
  csg_link, used in chunks 159c and 162.
Uses CSG_COMP and CsgNode 157a.
```

9.2.2 简单的 CSG 表达式语言

对于 CSG 表达式，此处将采用简单的语法。语言的元素则是图元对象、CSG 和分组运算符。其中，图元对象包含以下格式：c {p1 p2... }。其中，c 表示标识图元类的字母；p2 表示图元参数数值。CSG 运算符则由并集字符|、交集字符&、差集字符/标识。表达式分组则采用括号加以实现。

【例 9.2】 （CSG 表达式）下列代码分别对应于以下表达式：两个球体的并集（半径为 4，球心为(1,1,1)；半径为 3，球心为(2,2,2)）与另一个球体的交集（半径为 2）。

```
( s{0 0 0 2} & ( s{1 1 1 4}| s{2 2 2 3} ) )
```

该简单 CSG 表达式的解释器采用 lex 和 yacc 工具加以实现。

其中，词法分析器通过 lex 生成，并始于下列规范：

```
158b <csg lex 158b> ≡
  D [0-9]
  S [-]
  %%
  [ \t\n] ;
  {S}?{D}*"."{D}+ |
  {S}?{D}+"."{D}* |
  {S}?{D}+               { yylval.dval = atof( yytext ); return NUM; }
                         { return yytext[0]; }
```

上述描述定义了两种符号分类，即数字（D）和负号（S）。下列符号模式分别对应于空格、数字和独立的字符。

语法分析器则由 yacc 生成，语法的终止符号表示为数字（NUM）和字母。非终止符表示为 prim_obj、csg_obj 和 bop。

```
159a <csgunion 159a>≡
  %union {
    char cval;
    double dval;
    CsgNode *nval;
  }
Uses CsgNode 157a.

159b <csgclasses 159b>≡
  %token <dval> NUM
  %type <cval> bop
  %type <nval> prim_obj csg_obj
```

语法由语法的生成规则构成，如下所示。

```
159c <csg grammar 159c>≡
  csg_obj: '(' csg_obj bop csg_obj ')'  {$$ = root = csg_link($3, $2, $4);}
         | prim_obj
         ;
  bop:     '|'                          {$$ = '+'; }
         | '&'                          {$$ = '*'; }
         | '\\'                         {$$ = '-'; }
         ;
  prim_obj: 's' '{' NUM NUM NUM NUM '}' {$$ = csg_prim(sphere_set(
                                              sphere_instance(&sphere_funcs),
                                              v3_make($3, $4, $5), $6)); }
         ;
Uses csg_link 158a and csg_prim 157c.
```

注意，此处定义了单一图元类，即球体。其他图元可通过 prim_obj 添加至生成中。例程 csg_parse 解释 CSG 语言中的表达式，并调用 yacc、yyparse 生成的例程。

```
159d <csg parse 159d>≡
  CsgNode *csg_parse()
  {
    if (yyparse() == 0)
      return root;
    else
      return NULL;
  }
Defines:
  csg_parse, never used.
Uses CsgNode 157a.
```

9.2.3　三维场景描述语言中的 CSG 表达

针对 CSG 所选择的布尔表达式表示为适用于建模程序的表达，对应表达式较为简单，并能够捕捉该对象类的相关内容。然而，这一表达形式并不与三维场景描述语言兼容。对此，需要构建一个引擎，并在语言之间转换对象的描述。该转换过程始于 CSG 表达式树。

场景描述语言中的 CSG 对象表达遵循前缀表达式，并使用了下列运算符：csg_prim、csg_union、csg_inter 和 csg_diff，如下所示。

```
csg_union { csg_prim{ sphere { center = {0, 0, 0}}},
            csg_prim{ sphere { center = {1, 1, -1}}}
          }
```

例程 csg_write 将 CSG 对象写入其二叉树生成的文件中，并按深度递归遍历树形结构。当到达树的叶节点时，该例程利用 prim_write 例程写入其对应的图元（参见第 8 章）。在每个内部树节点中，该例程利用 csg_opname 例程写入对应 CSG 运算符的名称，并针对每棵子树执行递归操作。需要注意的是，上述处理类型（按深度遍历 CSG 树）可视为大多数 CSG 对象计算的基础操作。

```
160  <csg write 160>≡
 void csg_write(CsgNode *t, FILE *fd)
 {
   switch(t->type) {
   case CSG_PRIM: {
     fprintf(fd, "csg_prim{ "); prim_write(t->u.p, fd); fprintf(fd, " }\n");
     break; }
   case CSG_COMP:
     fprintf(fd, "%s {\n", csg_opname(t->u.c.op));
     csg_write(t->u.c.lft, fd); fprintf(fd, ",\n");
     csg_write(t->u.c.rgt, fd); fprintf(fd, "\n }");
     break;
   }
 }
Defines:
 csg_write, never used.
Uses CSG_COMP, csg_opname 161a, CSG_PRIM, csg_prim 157c, and CsgNode 157a.
```

例程 csg_opname 返回三维场景描述语言中的 CSG 运算符的名称字符串。

```
161a  <csg opename 161a>≡
  char *csg_opname(char c)
```

```
    {
     switch (c) {
     case '+': return "csg_union";
     case '*': return "csg_inter";
     case '-': return "csg_diff";
     default: return "";
     }
    }
Defines:
 csg_opname, used in chunk 160.
```

9.2.4 三维场景描述语言中的 CSG 对象的解释

除了写入三维场景描述语言中的 CSG 对象，还需要进一步解释该语言中的 CSG 表达式。对此，可实现与该语言中 CSG 运算符相关的函数即可，其余部分则通过解释器 lang_parse 完成。

例程 csg_prim_parse 通过将当前对象封装为 CSG 树中的叶节点来解释 CSG 图元。

```
161b <csg prim parse 161b>≡
  Val csg_prim_parse(int pass, Pval *p)
  {
    Val v;

    switch (pass) {
    case T_EXEC: {
      v.type = CSG_NODE;
      if (p != NULL && p->val.type == PRIM)
        v.u.v = csg_prim(p->val.u.v);
      else
        fprintf(stderr,"(csg_op): syntax error\n");
      break; }
    default:
    }
    return v;
  }
Defines:
 csg_prim_parse, never used.
Uses csg_prim 157c.
```

三维场景描述语言中的 CSG 运算符通过例程 csg_union_parse、csg_inter_parse 和 csg_diff_parse 予以解释。除了 CSG 操作代码之外，全部例程均包含相同的结构，因而此处仅显示 csg_union_parse 例程。

```
162 <csgunion parse 162>≡
 Val csg_union_parse(int pass, Pval *p)
 {
  Val v;
   switch (pass) {
   case T_EXEC: {
    if ((p ! = NULL && p->val.type == CSG_NODE)
        && (p->next ! = NULL && p->next->val.type == CSG_NODE)) {
      v.type = CSG_NODE;
      v.u.v = csg_link('+', p->val.u.v, p->next->val.u.v);
    } else {
      fprintf(stderr,"(csg_op): syntax error\n");
    }
    break; }
   default:
   }
   return v;
 }
Defines:
 csg_union_parse, never used.
Uses csg_link 158a.
```

注意，上述全部例程均使用了定义于 9.2 节中的 CSG 结构的构造函数。

9.3 生成式建模技术

生成式建模技术是使用转换分组和几何元素定义复杂形状的强大技术。

生成式模型 $g = (\gamma, \delta)$ 包含下列公式生成的参数函数描述：

$$S(u, v) = \delta(\gamma(u), v)$$

生成式模型的构成元素表示为生成器 γ 和转换组 δ。

其中，生成器通常为维度 1 的几何对象。也就是说，定义于环境空间内的参数曲线，如下所示。

$$\gamma(u) : \mathbb{R} \to \mathbb{R}^3$$

转换组定义了环境空间转换的单一参数族，如下所示。

$$\delta(p, v) : \mathbb{R}^3 \times \mathbb{R} \to \mathbb{R}^3$$

生成器对象的形状表示为生成器 $\gamma(u)$ 连续转换 $\delta(\cdot, v)$ 生成的参数表面 $S(u, v)$。需要注意的是，在该模型中，S 的参数表面包含了一个自然的分解：参数 u 是相对于当前生成器的；而 v 则参数化了转换组的动作。

兼具通用性和功能性的转换组定义为欧几里得空间的仿射转换组，该组通过下列方式定义：

$$h(p, v) = M_v(p) + T_v$$

其中，$p = (x, y, z)$ 表示为欧几里得空间点；$M_v(p) : \mathbb{R}^3 \to \mathbb{R}^3$ 表示为线性转换；而 $T_v \in \mathbb{R}^3$ 则定义为一个三维向量。注意，M_v 和 T_v 均取决于参数 v。

9.3.1 生成式模型的多边形近似表达

除了直接使用生成式模型的参数描述之外，还可与其几何近似表达结果协同工作。对此，可选择基于分解的表达形式，即对应于表面的分段式线性近似方法。

当构建生成式模型 (γ, δ) 给定的、与表面 $S(u, v)$ 近似的多边形网格时，可离散化其参数空间 $U = \{(u, v) : u \in [a, b], v \in [c, d]\}$。

U 的分解过程基于均匀采样，并对应于 $N \times M$ 点网格。在该网格的基础上，可利用 Coxeter-Freudenthal 分解法构建一个单体网格，如图 9.4 所示。

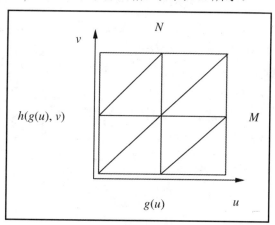

图 9.4 网格结构

需要注意的是，在水平方向上，存在一个沿曲线 $g(u)$ 的点采样，它被基于常量值 v 的 $h(\cdot, v)$ 函数所转换。

例程 gener_affine 生成分辨率为 $N \times M$ 的多边形网格，即 g 和 h 定义的生成式网格的近似结果。

```
164a <generaffine 164a>≡
  Poly *gener_affine(int n, Vector3 *g, int m, Matrix4 *h)
  {
    int u, v;
    Poly *tl = NULL;
    Vector3 *a = NEWARRAY(n, Vector3);
    Vector3 *b = NEWARRAY(n, Vector3);
    for (v = 0; v < m; v++) {
      for (u = 0; u < n; u++) {
        b[u] = v3_m4mult(g[u], h[v] );
        if (u == 0|| v == 0)
          continue;
        tl = poly_insert(tl, poly3_make(a[u-1], a[u], b[u-1] ));
        tl = poly_insert(tl, poly3_make(a[u], b[u], b[u-1] ));
      }
      SWAP(a, b, Vector3 *);
    }
    free(a); free(b);
    return tl;
  }
Defines:
  gener_affine, used in chunk 166a.
```

例程 affine_group 计算对应于转换分组离散化（由 t 中的规范定义）的矩阵数组，其中包含了转换代码和 p，后者包含了转换参数值。

```
164b <affine group 164b>≡
  Matrix4 *affine_group(int l, int m, char *t, Real **p)
  {
    int v;
    Matrix4 *h = NEWARRAY(m, Matrix4);
    for (v = 0; v < m; v++)
      h[v] = m4_compxform(l, t, p, v);
    return h;
  }
Defines:
  affine_group, used in chunk 166a.
Uses m4_compxform 164c.
```

例程 routine m4_compxform 计算基本转换连接后的转换矩阵。

```
164c <compxform 164c>≡
  Matrix4 m4_compxform(int k, char *t, Real **h, int j)
  {
```

```
    int i;
    Matrix4 m = m4_ident();

    for (i = 0; i < k; i++) {
      switch (t[i] ) {
        case G_TX: m = m4_m4prod(m4_translate(h[i] [j], 0, 0), m); break;
        case G_TY: m = m4_m4prod(m4_translate(0, h[i] [j], 0), m); break;
        case G_TZ: m = m4_m4prod(m4_translate(0, 0, h[i] [j] ), m); break;
        case G_RX: m = m4_m4prod(m4_rotate('x', h[i] [j] ), m); break;
        case G_RY: m = m4_m4prod(m4_rotate('y', h[i] [j] ), m); break;
        case G_RZ: m = m4_m4prod(m4_rotate('z', h[i] [j] ), m); break;
        case G_SX: m = m4_m4prod(m4_scale(h[i] [j], 1, 1), m); break;
        case G_SY: m = m4_m4prod(m4_scale(1, h[i] [j], 1), m); break;
        case G_SZ: m = m4_m4prod(m4_scale(1, 1, h[i] [j] ), m); break;
        default:
      }
    }
    return m;
  }
Defines:
  m4_compxform, used in chunk 164b.
Uses G_RX, G_RY, G_RZ, G_SX, G_SY, G_SZ, G_TX, G_TY, and G_TZ.
```

9.3.2 生成式模型的类型

生成式模型包含多种类型。其中较为常见的类型基于单一转换组，以及两个转换组的组合动作。

- 基于单一转换的生成式模型。
  - 挤压：平移操作。
  - 旋转：旋转操作。
- 基于两种转换的生成式模型。
  - 锥化：平移和缩放操作。
  - 弯曲：平移和旋转操作。
  - 扭转：平移和旋转操作。

注意，生成式模型并非唯一。例如，还可通过以下两种方式生成一个圆柱体。

- 圆形+挤压操作。
- 直线+旋转操作

9.3.3 旋转曲面

此处将通过示例展示生成式模型的实现过程。考查一个旋转曲面族，此类表面通过围绕该平面中某个轴的平面曲线的旋转操作生成。

通过曲线 g 的 360° 旋转，例程 rotsurf 生成旋转曲面的多边形近似结果。

```
166a <rotsurf 166a>≡
 Poly *rotsurf(int n, Vector3 *g, int m)
 {
   Matrix4 *h; Real *p[1]; Poly *s;
   char t[1] = {G_RY};
   p[0] = linear(0, PITIMES2, m);
   s = gener_affine(n, g, m, h = affine_group(1, m, t, p));
   efree(h);
   return s;
 }
Defines:
 rotsurf, never used.
Uses affine_group 164b, G_RY, gener_affine 164a, and linear 166b.
```

例程 linear 表示为一个辅助函数，并计算值 v_0 和 v_1 间的线性插值。

```
166b <linear 166b>≡
 Real *linear(Real v0, Real v1, int n)
 {
   int i;
   Real *x = NEWTARRAY(n, Real);
   Real incr = (v1 - v0) / (n -1);
   for (i = 0; i < n; i++)
     x[i] = v0 + (incr * i);
   return x;
 }
Defines:
 linear, used in chunk 166a.
```

9.4 补充材料

本章讨论了较为重要的建模技术及其实现库。

CSG 建模库的 API 由以下例程构成:

```
CsgNode *csg_parse();
CsgNode *csg_prim(Prim *p);
CsgNode *csg_link(int op, CsgNode *lft, CsgNode *rgt);

char *csg_opname(char c);
void csg_write(CsgNode *t, FILE *fd);

Val csg_union_parse(int c, Pval *p);
Val csg_inter_parse(int c, Pval *p);
Val csg_diff_parse(int c, Pval *p);
Val csg_prim_parse(int c, Pval *p);
```

生成式建模库的 API 由以下例程构成:

```
Poly *rotsurf(int n, Vector3 *g, int m);
Poly *gener_affine(int n, Vector3 *g, int m, Matrix4 *h);
Matrix4 m4_compxform(int k, char *t, Real **h, int j);
Matrix4 *affine_group(int l, int m, char *t, Real **p);

Real *linear(Real v0, Real v1, int n);
```

9.5 本章练习

（1）利用 CSG 库编写一个程序，并生成构建式实体模型。该程序应可接收形如（D = def）的几何图元定义。其中，D 表示为一个大写字母，def 表示图元定义。例如:

```
A = s {1, 1, 1, 4};
B = s {3, 2, 1, 10};

(A| (B & A))
```

（2）实现 csg_classify 操作，并相对于 CSG 模型执行点成员分类操作。

（3）利用 GENER 库编写一个程序并生成旋转表面。该程序应可读取一个二维表面，对应参数为旋转轴、旋转角和网格的离散化结果。

（4）在 CSG 表达中，实现几何转换操作。对此，需要定义 CSG_TRANSFORM 类型的 CSG 节点（Node）。

（5）利用 GP 和 CSG 库编写一个交互式程序，并创建 CSG 对象。该程序应支持图元、分组、转换（平移、旋转和缩放）和点成员操作（并集、差集、补集和交集）。

（6）编写一个程序，计算 CSG 模型的近似体积。

（7）利用 GP 和 GENER 库编写一个交互式程序，并生成旋转表面。该程序应支持曲线的输入和编辑操作，以及表面的绘制操作。

（8）实现基于挤压操作的生成式模型。

（9）编写两个程序创建圆柱体。① 使用旋转表面模型；② 使用挤压表面模型。

（10）实现生成式模型 TAPER、TWIST 和 BEND；编写一个程序，并根据此类模型创建表面。

第 10 章 层次结构和体系结构对象

在三维场景中，对象集合通常会保持一些物理关系，这也表明它们之前存在一定的几何关联。本章将讨论元素结构集的几何关系。

10.1 几何链接

几何转换构建了两个对象集合间的空间链接。在三维场景中，存在一个对每个对象都通用的坐标系，称为全局（或场景）坐标系。第 8 章曾讨论到，图元对象包含了一个局部坐标系。

这里，可将转换视为坐标系之间的转换。在这种情况下，转换可与对象集合关联，进而定义特定的坐标系。因此，我们将使用转换定义通用于对象集合的几何链接。共享该链接的对象隶属于同一个转换。

取决于链接类型，将包含不同的结构形式，其中较为常见的结构包括：

- ❑ 对象分组。转换用于定位空间内对象间的相对位置。一个特殊的例子是复合对象，并由子对象的并集构成（包含了子对象间的固定转换）。
- ❑ 关节式结构。该结构由几何链接（关节）关联的刚体（组件）构成。其中，转换包含两种方式：固定转换在关节坐标系间操作；变量转换对应于关节的自由度。

本章稍后将详细讨论这两种结构类型。

10.1.1 层次结构

某些对象集合由对象的子集构成，因而存在此类对象的层次结构。这一类结构可通过树形结构表达，其中包含了对应于独立对象 P_k 的叶节点，以及对应于对象 C_i 子集的内部节点。几何链接关系通常反映了此类结构的空间属性，因而也是层次结构表达中的一部分内容。相应地，可将转换 T_i 与每个子集 C_i 相关联，如图 10.1 所示。

在当前示例中，根据树形结构将以递归方式执行转换 T_i。换而言之，转换将影响集合中的元素，并传播至其后代。最终结果可描述为：转换 T_{P_k} 对应于每个独立的对象 P_k，并由对象所属的转换组合给出。

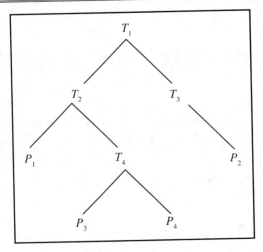

图 10.1 对象的层次结构

在许多计算机图形学应用程序中，可方便地与场景对象的非层次结构表达协同工作。对此，有必要将此类树形结构转换为一个列表，该转换过程称作平面化（flattening）。

非结构化表达表示为对象 (P_k, T_{P_k}) 及其转换形成的列表。当创建此类表达方式时，将计算组合转换 T_{P_k}，进而连接与每个层次结构级别关联的转换矩阵。这一处理过程可通过树形结构的深度遍历予以实现。

例如，图 10.1 所示的树形结构对应于下列列表：
$$((P_1, T_{P_1}), (P_2, T_{P_2}), (P_3, T_{P_3}), (P_4, T_{P_4}))$$

其中，组合转换由下列各式确定：
$$T_{P_1}(P_1) = T_1(T_2(P_1))$$
$$T_{P_2}(P_2) = T_1(T_2(T_4(P_3)))$$
$$T_{P_3}(P_3) = T_1(T_2(T_4(P_4)))$$
$$T_{P_4}(P_4) = T_1(T_3(P_1))$$

10.1.2 几何转换

与对象关联的转换应用需要使用到三维场景的全局坐标系和对象 P_k 本地坐标系间的映射，这可通过转换 T_k 给出，如图 10.2 所示。

在第 4 章中曾讨论到，针对参数化描述的对象，可直接使用转换 T；而对于隐式描述的对象，则使用逆转换 T^{-1}。

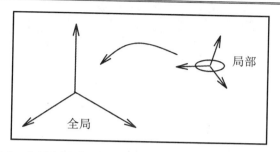

图 10.2　全局坐标系（场景）和局部坐标系（对象）间的转换

10.1.3　仿射不变性

本节将选择仿射转换类实现聚合和层次结构中的几何链接，其主要原因在于：除了包含全部重要的转换之外，该类还支持建模和视见处理过程中高效的转换应用。

在这一上下文环境中，该几何转换类应具有仿射不变性。当给定图形对象 O 的离散化结果 D_O 以及转换 T 后，这一属性可确保 T 和 D_O 元素的操作结果等同于重构 O 并应用 T 后的结果。如图 10.3 所示的交换图显示了这一属性，其中，R 表示为重构运算符。

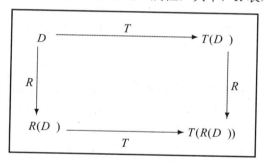

图 10.3　交换图

下面通过一个具体的例子以更好地理解此属性的重要性。

【例 10.1】　（直线）考查参数表达 $g:[0,1] \to \mathbb{R}^3$ 给出的直线段 $\overline{p_0 p_1}$，其中 $g(u) = u p_0 + (1-u) p_1$，且有 $p_0, p_1 \in \mathbb{R}^3$。在该示例中，重构运算符表示为 $R(p_0, p_1) \equiv g$。

如果 T 表示为仿射转换，则有：

$$T(R(p_0, p_1)(u)) = R(T(p_0), T(p_1))(u)$$

或

$$T(g(u)) = u T(p_0) + (1-u) T(p_1)$$

该属性最大的优点是可以转换直线段，向端点 p_0 和 p_1 应用 T 已然足够，如图 10.4 所示。

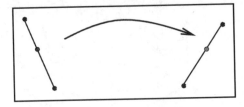

图 10.4 仿射不变性和转换

上述示例结果对于大多数图形对象的几何描述均为有效。
- ❏ 直线和多边形（顶点）。
- ❏ 曲线和参数表面（控制点）[①]。
- ❏ 隐式表面（几何元素）。

10.2 层次结构和转换

本节将讨论与层次结构对象关联的转换计算实现。对应操作实际上涉及表达层次结构的树形节点的深度访问。也就是说，从树根节点至叶节点，沿着下降路径在每个内部节点上执行转换组合操作。在每个叶节点中，组合转换应用于相关对象上。相应地，这一处理过程可采用递归方式加以实现。需要注意的是，应在每个树形节点处保留组合变换的中间结果。

当实现上述操作时，栈是较为适宜的数据结构，进而可高效地存储计算过程的部分数据。另外，转换操作定义为一个 4×4 矩阵。在处理过程的每一个步骤中，当前转换矩阵（CTM）则对应于栈顶。当实现栈引擎时，可定义相应的压栈和弹栈操作，并以此访问当前数据结构。除此之外，这里还将定义基于当前转换矩阵的操作。

10.2.1 栈操作

结构 Stack4 定义了一个栈，其中包含了最大尺寸（size），以及基于直接和逆转换（mbot 和 ibot）的数组。此外，指针 mtop 和 itop 指向栈顶。

```
173a <stack 4 173a>≡
  typedef struct Stack4 {
    int size;
    Matrix4
```

[①] NURBS（非均匀有理 B 样条）也具有投影不变性。

第 10 章 层次结构和体系结构对象

```
      *mbot, *mtop,
      *ibot, *itop;
  } Stack4;
Defines:
  Stack4, used in chunks 173-76.
```

例程 s4_initstac 定义为栈构造函数。

```
173b <initstack 173b>≡
  Stack4 *s4_initstack(int size)
  {
    int i;
    Matrix4 *m;
    Stack4 *s = (Stack4 *) emalloc(sizeof (Stack4));
    s->size = size;
    s->mbot = ( Matrix4 * ) emalloc( size * sizeof(Matrix4));
    s->ibot = ( Matrix4 * ) emalloc( size * sizeof(Matrix4));

    for (m = s->mbot, i = 0; i < s->size; i++)
      *m++ = m4_ident();
    for (m = s->ibot, i = 0; i < s->size; i++)
      *m++ = m4_ident();

    s->mtop = s->mbot;
    s->itop = s->ibot;

    return s;
  }
Defines:
  s4_initstack, used in chunk 179a.
Uses Stack4 173a.
```

例程 s4_push 将当前转换压入栈中。

```
174a <push 174a>≡
  void s4_push(Stack4 *s)
  {
    Matrix4 *m;
    if ((s->mtop - s->mbot) >= (s->size - 1))
      error("(s4_push): stack overflow");
    m = s->mtop;
    s->mtop++;
    *s->mtop = *m;
    m = s->itop;
```

```
    s->itop++;
    *s->itop = *m;
    }
Defines:
  s4_push, used in chunk 179a.
Uses Stack4 173a.
```

例程 s4_pop 将当前转换弹栈。

```
174b <pop 174b>≡
  void s4_pop(Stack4 *s)
  {
    if (s->mtop <= s->mbot)
      error("(s4_pop()) stack underflow\n");
    s->mtop--;
    s->itop--;
  }
Defines:
  s4_pop, used in chunk 179a.
Uses Stack4 173a.
```

10.2.2 转换

这里将构造构成平移、旋转和缩放基本转换的仿射转换。

例程 s4_translate 将转换矩阵与栈顶连接起来。

```
174c <stranslate 174c>≡
  void s4_translate(Stack4 *s, Vector3 t)
  {
    *s->mtop = m4_m4prod(*s->mtop, m4_translate( t.x, t.y, t.z));
    *s->itop = m4_m4prod(m4_translate(-t.x,-t.y,-t.z), *s->itop);
  }
Defines:
  s4_translate, used in chunk 179b.
Uses Stack4 173a.
```

例程 s4_scale 将缩放矩阵和栈顶连接起来。

```
175a <sscale 175a>≡
  void s4_scale(Stack4 *s, Vector3 v)
  {
    *s->mtop = m4_m4prod(*s->mtop, m4_scale(v.x, v.y, v.z));
```

第 10 章 层次结构和体系结构对象

```
    if (REL_EQ(v.x, 0.0)|| REL_EQ(v.y,0.0)|| REL_EQ(v.z,0.0))
      fprintf(stderr,"(s4_scale()) unable to invert scale matrix\n");
    else
      *s->itop = m4_m4prod(m4_scale(1./v.x,1./v.y,1./v.z), *s->itop);
  }
Defines:
  s4_scale, used in chunk 180a.
Uses Stack4 173a.
```

例程 **s4_rotate** 将旋转矩阵和栈顶连接起来。

```
175b <srotate 175b>≡
  void s4_rotate(Stack4 *s, char axis, Real angle)
  {
    *s->mtop = m4_m4prod(*s->mtop, m4_rotate(axis,angle));
    *s->itop = m4_m4prod(m4_rotate(axis,-angle), *s->itop);
  }
Defines:
  s4_rotate, used in chunk 180b.
Uses Stack4 173a.
```

例程 **s4_v3xform** 和 **s4_n3xform** 向几何元素应用当前转换。例程 **s4_v3xform** 将转换一个向量。

```
175c <v3xform 175c>≡
  Vector3 s4_v3xform(Stack4 *s, Vector3 v)
  {
    return v3_m4mult(v, *s->mtop);
  }
Defines:
  s4_v3xform, never used.
Uses Stack4 173a.
```

例程 **s4_v3xform** 转换切平面的法线方向。

```
175d <n3xform 175d>≡
  Vector3 s4_n3xform(Stack4 *s, Vector3 nv)
  {
    return v3_m4mult(nv, m4_transpose(*s->itop));
  }
Defines:
  s4_n3xform, never used.
Uses Stack4 173a.
```

例程 s4_getmat 和 s4_getimat 分别返回当前的直接和逆转换。

```
176a  <getmat 176a>≡
  Matrix4 s4_getmat(Stack4 *s)
  {
    return *s->mtop;
  }
Defines:
  s4_getmat, used in chunk 181a.
Uses Stack4 173a.
```

```
176b  <getimat 176b>≡
  Matrix4 s4_getimat(Stack4 *s)
  {
    return *s->itop;
  }
Defines:
  s4_getimat, used in chunk 181a.
Uses Stack4 173a.
```

例程 s4_loadmat 可用于调整当前转换矩阵。

```
176c  <loadmat 176c>≡
  void s4_loadmat(Stack4 *s, Matrix4 *md, Matrix4 *im)
  {
    *s->mtop = *md;
    *s->itop = (im == (Matrix4 *)0)? m4_inverse(*md) : *im;
  }
Defines:
  s4_loadmat, never used.
Uses m4_inverse and Stack4 173a.
```

例程 s4_concmat 将某个矩阵连接至当前转换中。

```
176d  <concmat 176d>≡
  void s4_concmat(Stack4 *s, Matrix4 *md, Matrix4 *im)
  {
    *s->mtop = m4_m4prod(*md, *s->mtop );
    if ( im == (Matrix4 *)0)
      *s->itop = m4_m4prod(*s->itop, m4_inverse(*md));
    else
      *s->itop = m4_m4prod(*s->itop, *im);
  }
Defines:
```

```
s4_concmat, never used.
Uses m4_inverse and Stack4 173a.
```

10.3 对象分组

针对对象的层次结构分组,本节将讨论三维场景描述语言的实现。该描述适用于组合和关节式对象。

10.3.1 层次结构描述

对象层次结构所采用的格式基于 group 结构,其中包含了转换(平移、旋转和缩放)和对象的子分组(children)。

```
177a <group scn 177a>≡
 group{
     transform = { translate {v = {0, .0, 0}}, rotate {z = 0 }},
     children = group {
                 transform = { translate {v = {.1, 0, 0}}},
                 children = primobj{ shape = sphere{radius = .1 }}
             },
     transform = { translate {v = {.2, 0, 0}}, rotate {z = 0 }},
     children = group {
                 transform = { translate {v = {.2, 0, 0}}},
                 children = primobj{ shape = sphere{radius = .1}}
             }
 };
```

10.3.2 对象

在前述与建模机制相关的章节中,曾针对几何对象介绍了多种表达模式。在定义对象分组前,需要准确地定义系统中的对象概念。这里,对象由结构 Object 表述,其中包含了几何支持(shape)及其属性,如材质类型。

```
177b <object struct 177b>≡
 typedef struct Object {
   struct Object  *next;
   struct Material *mat;
   int type;  /* shape */
```

```
    union {
      struct Poly    *pols;
      struct Prim    *prim;
      struct CsgNode *tcsg;
    } u;
  } Object;
Defines:
  Object, used in chunks 178, 181, 182, and 187a.
```

需要注意的是，上述结构的目的是将应用于系统的不同几何表达模式封装在一个计算实体中。通过这种方式，对象的形式可采用基于图元、分解模式（多边形网格）、构建模式（CSG）的表述。

```
#define V_CSG_NODE   901
#define V_PRIM       902
#define V_POLYLIST   903
```

注意，结构 Object 设计为对象单链表中的元素，并可用于表示对象集合。

例程 obj_new 定义为对象构造函数，该函数将分配结构，并初始化对应于对象几何体的字段。宏 SET_MAT_DEFAULT 将于稍后加以定义。

```
178a <obj new 178a>≡
  Object *obj_new(int type, void *v)
  {
    Object *o = NEWSTRUCT(Object);
    o->next = NULL;
    SET_MAT_DEFAULT(o);
    switch (o->type = type) {
    case V_CSG_NODE: o->u.tcsg = v; break;
    case V_PRIM: o->u.prim = v; break;
    case V_POLYLIST: o->u.pols = v; break;
    default: error("(newobj) wrong type");
    }
    return o;
  }
Defines:
  obj_new, never used.
Uses Object 177b and SET_MAT_DEFAULT.
```

例程 obj_free 定义为对象的析构函数。

```
178b <obj free 178b>≡
  void obj_free(Object *o)
```

```
  switch (o->type) {
  case V_PRIM: prim_destroy(o->u.prim); break;
  case V_CSG_NODE: csg_destroy(o->u.tcsg); break;
  case V_POLYLIST: plist_free(o->u.pols); break;
  }
  efree(o->mat->tinfo); efree(o->mat);
  efree(o);
}
Defines:
  obj_free, used in chunk 182c.
Uses Object 177b.
```

10.3.3 分组和对象列表

这里将针对链接对象采用两种描述类型。从外部来看，将使用 10.3.1 节定义的层次结构描述；从内部来看，则使用基于对象列表的等效非层次结构描述。

外部和内部表达间的转换将通过三维场景描述语言予以执行，稍后将对此加以讨论。处理层次结构的例程将基于栈引擎，并使用了静态结构 Stack4。

```
static Stack4 *stk = NULL;
```

例程 group_parse 执行层次结构树的深度访问，该例程将实现三维场景描述语言中运算符 group 的语义，并按照下降方向，即朝向树形结构的叶节点压入当前转换，同时按照上升方向，即朝向树形结构的根节点弹出转换。在各节点处，将执行层次结构对应级别的对象转换操作，调用 transform_objects 例程，并利用 routine collect_objects 创建包含子树对象的列表。稍后将展示这两个例程。

```
179a <group parse 179a>≡
  Val group_parse(int pass, Pval *pl)
  {
    Val v = {V_NULL, 0};
    switch (pass) {
    case T_PREP:
      if (stk == NULL) stk = s4_initstack(MAX_STK_DEPTH);
      s4_push(stk);
      break;
    case T_EXEC:
      transform_objects(pl);
      s4_pop(stk);
```

```
      v.u.v = collect_objects(pl); v.type = V_GROUP;
      break;
    }
    return v;
  }
Defines:
  group_parse, never used.
Uses collect_objects 181c, MAX_STK_DEPTH, s4_initstack 173b, s4_pop 174b,
  s4_push 174a, stk, and transform_objects 181a.
```

例程 translate_parse、scale_parse 和 rotate_parse 实现了当前语言中转换运算符的语义；此外还利用当前转换矩阵执行了特定转换的连接操作。

```
179b <translate parse 179b>≡
  Val translate_parse(int pass, Pval *p)
  {
    Val v = {V_NULL, 0};
    if (pass == T_EXEC) {
      if (p->val.type == V_PVL)
        s4_translate(stk, pvl_to_v3(p->val.u.v));
      else
        error("(translate) wrong argument");
    }
    return v;
  }
Defines:
  translate_parse, never used.
Uses s4_translate 174c and stk.

180a <scale parse 180a>≡
  Val scale_parse(int pass, Pval *p)
  {
    Val v = {V_NULL, 0};
    if (pass == T_EXEC) {
      if (p->val.type == V_PVL)
        s4_scale(stk, pvl_to_v3(p->val.u.v));
      else
        error("(scale) wrong argument");
    }
    return v;
  }
Defines:
  scale_parse, never used.
```

Uses s4_scale 175a and stk.

180b <rotate parse 180b>≡
```
  Val rotate_parse(int pass, Pval *p)
  {
    Val v = {V_NULL, 0};
    if (pass == T_EXEC) {
      if (strcmp(p->name, "x") == 0 && p->val.type == V_NUM )
        s4_rotate(stk, 'x', p->val.u.d);
      else if (strcmp(p->name, "y") == 0 && p->val.type == V_NUM )
        s4_rotate(stk, 'y', p->val.u.d);
      else if (strcmp(p->name, "z") == 0 && p->val.type == V_NUM )
        s4_rotate(stk, 'z', p->val.u.d);
      else
        error("(rotate) wrong argument");
    }
    return v;
  }
```
Defines:
 rotate_parse, never used.
Uses s4_rotate 175b and stk.

10.3.4 对象转换

例程 transform_objects 遍历隶属于某个分组的对象列表,并将当前转换应用于其中的每个对象上。

181a <transform objects 181a>≡
```
  static void transform_objects(Pval *pl)
  {
    Pval *p;
    for (p = pl; p != NULL; p = p->next)
      if (p->val.type == V_OBJECT)
        obj_transform(p->val.u.v, s4_getmat(stk), s4_getimat(stk));
  }
```
Defines:
 transform_objects, used in chunk 179a.
Uses obj_transform 181b, s4_getimat 176b, s4_getmat 176a, and stk.

例程 obj_transform 根据对应的类型转换某个对象。

181b <obj xform 181b>≡
```
  void obj_transform(Object *o, Matrix4 m, Matrix4 mi)
```

```
  switch (o->type) {
  case V_PRIM: prim_transform(o->u.prim, m, mi); break;
  case V_CSG_NODE: csg_transform(o->u.tcsg, m, mi); break;
  case V_POLYLIST: plist_transform(o->u.pols, m); break;
  }
}
Defines:
 obj_transform, used in chunk 181a.
Uses Object 177b.
```

10.3.5 收集列表中的对象

例程 group_parse 作为参数接收一个列表，其中包含了各自对象以及对象列表。这一类对象表示为定义于当前分组范围内的图元；对象列表则表示为该对象分组的子分组。

例程 collect_objects 利用某个分组的所有隶属对象创建一个集合。换而言之，该过程将层次结构中特定节点的子树元素置于单一列表中。

```
181c <collect objects 181c>≡
 static Object *collect_objects(Pval *pl)
 {
  Pval *p; Object *olist = NULL;
  for (p = pl; p != NULL; p = p->next) {
   if (p->val.type == V_OBJECT)
     olist = obj_insert(olist, p->val.u.v);
    else if (p->val.type == V_GROUP)
     olist = obj_list_insert(olist, p->val.u.v);
  }
  return olist;
 }
Defines:
 collect_objects, used in chunk 179a.
Uses obj_insert 182a, obj_list_insert 182b, and Object 177b.
```

例程 obj_insert 将对象插入列表中。

```
182a <obj insert 182a>≡
 Object *obj_insert(Object *olist, Object *o)
 {
  o->next = olist;
  return o;
 }
```

例程 obj_list_insert 将一个对象列表插入列表中。

```
182b <obj list insert 182b>≡
  Object *obj_list_insert(Object *olist, Object *l)
  {
    Object *t, *o = l;
    while (o ! = NULL) {
      t = o; o = o->next;
      olist = obj_insert(olist, t);
    }
    return olist;
  }
Defines:
  obj_list_insert, used in chunk 181c.
Uses obj_insert 182a and Object 177b.
```

例程 obj_list_free 释放分配给对象列表的内存空间。

```
182c <obj list free 182c>≡
  void obj_list_free(Object *ol)
  {
    Object *t, *o = ol;
    while (o ! = NULL) {
      t = o; o = o->next;
      obj_free(t);
    }
  }
Defines:
  obj_list_free, used in chunk 188b.
Uses obj_free 178b and Object 177b.
```

10.3.6 参数化链接

关节式结构由可变的几何链接组成,并由依赖关系图定义。该图包含以下元素。

❑ 链接:坐标系间的映射机制。
❑ 关节:可变的转换(自由度)。
❑ 对象:局部几何体。

【例 10.2】 （关节式胳膊对象）。

```
183a <arm 183a>≡
  group {
    transform = { translate {v = {0, 0, 0}},
                  rotate {z = arg{ r1 = 0 }}},
    children = group { children = primobj{ shape = cylinder { height = 1 }}},
    transform = { translate {v = {1, 0, 0}},
                  rotate {z = motor{ arg{ r2 = 0 }}}},
    children = group { children = primobj{ shape = cylinder { height = 1 }}},
  }
```

与关节式结构的自由度关联的转换值可通过多种方式获得。一种实现方式是使用定义于三维场景描述语言解释器命令行中的变量，稍后将对此予以解释。

例程 arg_init 利用源自程序命令行中的参数列表初始化局部结构。

```
static int m_argc;
static char **m_argv;

183b <arginit 183b>≡
  void arg_init(int ac, char **av)
  {
    m_argc = ac; m_argv = av;
  }
Defines:
  arg_init, never used.
Uses m_argc and m_argv.
```

例程 arg_parse 实现了三维场景描述语言中的运算符 arg。

```
183c <arg parse 183c>≡
  Val arg_parse(int pass, Pval *p)
  {
    Val v = {V_NULL, 0};
    switch (pass) {
    case T_EXEC:
      if (p != NULL && p->val.type == V_NUM)
        v.u.d = arg_get_dval(p->name, p->val.u.d);
      else
        fprintf(stderr, "error: arg parse %lx\n",p);
      v.type = V_NUM;
      break;
    }
```

```
    return v;
  }
Defines:
  arg_parse, never used.
Uses arg_get_dval 184.
```

例程 arg_get_dval 在参数列表中搜索类型-值对。若存在,则返回 value;否则返回 defval。

```
184 <arg_get_dval 184>≡
  double arg_get_dval(char *s, Real defval)
  {
    int i;
    for (i = 1; i < m_argc; i++)
      if (m_argv[i] [0] == '-' && strcmp(m_argv[i] +1, s) == 0 && i+1 < m_argc)
        return atof(m_argv[i+1] );
    return defval;
  }
Defines:
  arg_get_dval, used in chunk 183c.
Uses m_argc and m_argv.
```

需要注意的是,上述例程适用于多种上下文环境中。

10.4 动　　画

本节将讨论一种计算模式,以支持通用动画的过程式结构,特别是关节式对象。

10.4.1 动画时钟

当实现过程式动画时,需要定义相关例程进而对时钟加以控制。相应地,当前动画时间通过变量 time 加以定义;布尔变量 stop 则表示时钟是否处于停止状态。

```
static Real time = 0;
static Boolean stop = FALSE;
```

例程 time_reset 将重启时钟。

```
185a <time reset 185a>≡
  void time_reset(Real t)
  {
    time = t;
```

```
      stop = FALSE;
    }
Defines:
   time_reset, used in chunk 187b.
Uses stop and time.
```

例程 time_done 表示动画时间是否结束。

```
185b <time done 185b>≡
   Boolean time_done(Real tlimit)
    {
      return (time > tlimit| | stop == TRUE);
    }
Defines:
   time_done, used in chunk 187b.
Uses stop and time.
```

例程 time_incr 将增加时间。

```
185c <time incr 185c>≡
   Real time_incr(Real tincr)
    {
      if (! stop)
        time += tincr;
      return time;
    }
Defines:
   time_incr, used in chunk 187b.
Uses stop and time.
```

例程 time_get 返回当前时间。

```
185d <time get 185d>≡
   Real time_get()
    {
      return time;
    }
Defines:
   time_get, used in chunk 186.
Uses time.
```

例程 time_end 将终止时钟并结束动画。

```
185e <time end 185e>≡
   Real time_end()
```

```
    {
      stop = TRUE;
      return time;
    }
Defines:
  time_end, never used.
Uses stop and time.
```

10.4.2 过程式动画的构建

下面将通过实例考查动画的过程式构建。

【例 10.3】 （引擎）。动画运算符实现了基于恒定速度的引擎，对应语法为 engine {IN A}。

```
transform = { rotate { z = motor{.2 }}}
```

例程 motor_parse 实现了运算符 engine，同时还计算了当前时间的对应值。

```
186 <motor parse 186>≡
  Val motor_parse(int pass, Pval *p)
  {
    Val v = {V_NULL, 0};
    switch (pass) {
    case T_EXEC:
      if (p != NULL && p->val.type == V_NUM)
        v.u.d = time_get() * p->val.u.d;
      else
        fprintf(stderr, "error: motor parse\n");
      v.type = V_NUM;
      pvl_free(p);
      break;
    }
    return v;
  }
Defines:
  motor_parse, never used.
Uses time_get 185d.
```

10.4.3 动画的执行过程

当采用过程式动画结构时，我们有一种方法可确定多个参数值，这些参数随着与三维场景中的对象相关的时间而变化。

针对于此，将定义一个新的数据结构 Scene，其中包含了所有的场景对象列表。稍后还将引入该结构中的其他元素。

```
187a <scene structure 187a>≡
  typedef struct Scene {
    struct Object *objs;
  } Scene;
Defines:
  Scene, used in chunks 187 and 188.
Uses Object 177b and objs.
```

当可视化动画效果时，需要在当前时刻生成场景的图像序列。

下列程序实现了动画帧序列的生成过程。

```
187b <anim 187b>≡
  int main(int argc, char **argv)
  {
    Scene *s;
    init_scene();
    time_reset(0);
    s = scene_eval(scene_read());
    while (! time_done(timeoff)) {
      render_frame(s, get_time());
      scene_free(s);
      s = scene_eval();
      time_incr(1);
    }
  }
Defines:
  main, used in chunks 313, 317, and 318c.
Uses Scene 187a, scene_eval 188a, scene_free 188b, scene_read 187c,
time+done 185b, time_incr 185c, and time_reset 185a.
```

例程 scene_read 读取包含三维场景描述和动画结构的文件。

```
187c <scene read 187c>≡
  Scene *scene_read(void)
  {
    if (lang_parse() == 0)
      return lang_ptree();
    else
      error("(scene read)");
```

```
}
Defines:
  scene_read, used in chunk 187b.
Uses Scene 187a.
```

例程 scene_eval 解释当前时刻 t 时的三维场景描述和动画结构。需要注意的是，该例程使用了当前场景描述的无损型评估（lang_nd_eval）。

```
188a <scene eval 188a>≡
  Scene *scene_eval(void)
  {
    Scene *s;
    Val v = lang_nd_eval();
    if (v.type ! = V_SCENE)
      error("(scene eval)");
    else
      s = v.u.v;
    return s;
  }
Defines:
  scene_eval, used in chunk 187b.
Uses Scene 187a.
```

例程 scene_free 释放时刻 t 时三维场景所分配的内存空间。

```
188b <scene free 188b>≡
  void scene_free(Scene *s)
  {
    if (s->objs)
      obj_list_free(s->objs);
    efree(s);
  }
Defines:
  scene_free, used in chunk 187b.
Uses obj_list_free 182c, objs, and Scene 187a.
```

10.5 补 充 材 料

本章讨论了动画中的层次结构及其应用，并针对层次结构和过程式动画开发了相关库。

层次结构库中的 API 涵盖了以下例程。

```
Stack4 *s4_initstack(int size);
void s4_push(Stack4 *s);
void s4_pop(Stack4 *s);

void s4_translate(Stack4 *s, Vector3 t);
void s4_scale(Stack4 *s, Vector3 v);
void s4_rotate(Stack4 *s, char axis, Real angle);

Vector3 s4_v3xform(Stack4 *s, Vector3 v);
Vector3 s4_n3xform(Stack4 *s, Vector3 nv);

Matrix4 s4_getmat(Stack4 *s);
Matrix4 s4_getimat(Stack4 *s);
void s4_loadmat(Stack4 *s, Matrix4 *md, Matrix4 *im);
void s4_concmat(Stack4 *s, Matrix4 *md, Matrix4 *im);
```

实现了场景描述语言中层次结构的例程包括：

```
Val group_parse(int pass, Pval *pl);
Val translate_parse(int pass, Pval *p);
Val scale_parse(int pass, Pval *p);
Val rotate_parse(int pass, Pval *p);
```

动画库中的 API 涵盖以下例程：

```
void time_reset(Real t);
Boolean time_done(Real tlimit);
Real time_incr(Real tincr);
Real time_get();
Real time_end();

Val motor_parse(int pass, Pval *p);

void arg_init(int ac, char **av);
double arg_get_dval(char *s, Real de
Val arg_parse(int pass, Pval *p);
```

10.6　本章练习

（1）编写一个交互式程序，将转换与图元对象进行关联。

（2）编写一个程序，解释并可视化层次结构对象，该对象由图元构成，并在 SDL 语言中予以描述。相应地，可使用 arg 命令将参数传递至程序的命令行中。

（3）修改练习（1）和（2）中的程序，以使其可与 CSG 对象协同工作。

（4）修改练习（1）和（2）中的程序，以使其可与多边形网格协同工作。

（5）编写一个程序，并显示窗口中的模拟时钟。

（6）编写一个程序，并采用交互方式创建一个关节式序列。

（7）编写一个程序，扫描关节式序列，并对其实现动画效果。

（8）采用 SDl 语言编写一个关节式角色的描述。

第 11 章 视见相机转换

三维场景的视见机制需要将三维数据从世界空间中转换至图像中的二维动画。其间，相机转换则是该处理过程中的一个主要组件。前述章节曾讨论了三维对象建模的转换应用，本章将介绍视见转换。

11.1 视见处理过程

视见处理涵盖了一系列的操作，同时涉及将场景的三维对象映射至图像的二维投影中。

11.1.1 视见操作和参考空间

视见操作包含以下内容：
- 相机的映射机制。
- 剔除和剪裁。
- 透视转换。
- 可见性计算。
- 光栅化操作。

其中，各项操作均包含了不同的特征。对此，如果选取了适宜的坐标系，则可通过简单、高效的方式予以实现。视见处理（或视见管线）对应于一个转换和操作序列；随后，场景对象将依次映射至参考空间内，并于其中执行视见操作，如图 11.1 所示。针对视见操作，参考空间包含以下内容：
- 对象（或模型）空间。
- 世界（或场景）空间。
- 相机空间。
- 可见空间。
- 图像空间。

（1）对象空间。对象空间使用了对象的局部坐标系。该空间包含以下特征：原点对应于对象的质心，主方向之一与该对象的最大轴对齐。另外，各维度均处于标准化状态。对象的建模操作和几何计算在该空间内执行。

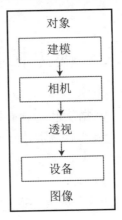

图 11.1 视见转换序列

（2）世界空间。世界空间使用应用程序的全局坐标系，同时也是各个场景对象的公共坐标系，其维度采用与应用程序相关的单位加以定义。另外，可在该空间内执行光照计算。

（3）相机空间。相机空间使用了观察者的坐标系，视见方向对应于 z 轴，图像平面平行于 xy 平面。视见体映射于标准化的金字塔空间中。相应地，可在该空间内执行剪裁操作。

（4）可见空间。可见空间使用了源自投影转换的坐标系，其中，投影中心映射至理想点上。通过这种方式，视见体变为一个平行六面体。在该空间内，可执行可见表面的计算。

（5）图像空间。图像空间使用了图形设备的坐标系，并包含了离散坐标。在该空间内，可执行光栅化操作。

11.1.2　虚拟相机和视见参数

视见规范的执行方式遵循于虚拟相机模型。该模型定义了必要的参数，以生成三维场景图像。视见参数包含以下内容：

- 投影中心。
- 视见方向。
- 垂直的视见上向量。
- 投影的视见平面。
- 前、后平面。
- 图像的中心和维度。

第 11 章 视见相机转换

❑ 投影类型。

数据结构 View 定义了视见对象，其中包含了全部视见参数及其对应的转换操作。

```
193 <view structure 193>≡
  typedef struct View {
    Vector3 center;           /* center of projection */
    Vector3 normal;           /* view plane normal vector */
    Vector3 up;               /* view up vector */

    Real dist;                /* view plane distance from viewpoint*/
    Real front;               /* front plane dist.from viewpoint */
    Real back;                /* back plane dist.from viewpoint */

    UVpoint c;                /* relative to view plane center */
    UVpoint s;                /* window u,v half sizes */

    Box3d sc;                 /* current, in pix space */

    int type;                 /* projection type */

    Matrix4 V, Vinv;          /* view xform and inverse */
    Matrix4 C, Cinv;          /* clip space xform and inverse */
    Matrix4 P, Pinv;          /* perspective xforms and inverse */
    Matrix4 S, Sinv;          /* device xform and inverse */
  } View;
Defines:
  View, used in chunks 194c, 195a, and 206-9.
Uses perspective 207b, UVpoint 194a, and view.
```

投影类型可指定为透视（圆锥形）或正交（平行）。

```
#define PERSPECTIVE 1
#define ORTHOGRAPHIC 2
```

UVpoint 结构表示为图像平面上的一个向量。

```
194a <uv point 194a>≡
  typedef struct UVpoint {
    Real u,v;
  } UVpoint;
Defines:
  UVpoint, used in chunk 193.
```

下列宏定义了一些有用的关系。

```
194b <view relations 194b>≡
 zmin = view.front / view.back
 AspectRatio = view.s.u / view.s.v
 fieldofView = 2 * arctan(view.s.u / view.dist) n
 PixelAspect = aspect * ((sc_upper.y - sc_lower.y +1) / (sc_upper.x -
                                                         sc_lower.x +1))
Uses view.
```

11.1.3 定义视见参数

访问 View 结构时，需要通过一组例程完成，进而指定相应的事件参数。此类例程负责验证数据的一致性，确保该结构展现一个有效的视见体，并访问一个指向视见对象（存储于内部变量 view 中）的指针。

```
static View *view;
```

例程 setview 初始化变量 view。

```
194c <setview 194c>≡
 void setview(View *v)
 {
   view = v;
 }
Defines:
 setview, used in chunks 206-8.
Uses View 193 and view.
```

例程 getview 返回当前视见对象。

```
195a <getview 195a>≡
 View *getview(void)
 {
   return view;
 }
Defines:
 getview, used in chunk 206a.
Uses View 193 and view.
```

例程 setviewpoint 定义对应的视点。

```
195b <setviewpoint 195b>≡
 void setviewpoint(Real x, Real y, Real z)
```

第 11 章 视见相机转换

```
  {
    view->center.x = x;
    view->center.y = y;
    view->center.z = z;
  }
Defines:
  setviewpoint, used in chunks 197c, 206b, 207a, and 209.
Uses view.
```

例程 setviewnormal 定义了视见方向，并对其执行标准化操作。

```
195c <setviewnormal 195c>≡
  void setviewnormal(Real x, Real y, Real z)
  {
    double d = sqrt(SQR(x)+SQR(y)+SQR(z));

    if (d < ROUNDOFF)
      error("invalid view plane normal");
    view->normal.x = x / d;
    view->normal.y = y / d;
    view->normal.z = z / d;
  }
Defines:
  setviewnormal, used in chunks 197c, 206b, 207a, and 209.
Uses ROUNDOFF and view.
```

例程 setviewup 定义了视见的上向量。

```
195d <setviewup 195d>≡
  void setviewup(Real x, Real y, Real z)
  {
    if (fabs(x) + fabs(y) + fabs(z) < ROUNDOFF)
      error("no view up direction");
    view->up.x = x;
    view->up.y = y;
    view->up.z = z;
  }
Defines:
  setviewup, used in chunks 197c, 206b, and 207a.
Uses ROUNDOFF and view.
```

例程 setviewdistance 定义了投影平面，也就是说，在视见方向上始于投影中心的距离。

```
196a <setviewdistance 196a>≡
  void setviewdistance(Real d)
```

```
    {
      if (fabs(d) < ROUNDOFF)
        error("invalid view distance");
      view->dist = d;
    }
Defines:
  setviewdistance, used in chunks 197c, 207, and 208a.
Uses ROUNDOFF and view.
```

例程 setviewdepth 定义了剪裁体的近平面和远平面。

```
196b <setviewdepth 196b>≡
    void setviewdepth(Real front, Real back)
    {
      if (fabs(back - front) < ROUNDOFF|| fabs(back) < ROUNDOFF)
        error("invalid viewdepth");
      view->front = front;
      view->back = back;
    }
Defines:
  setviewdepth, used in chunks 197c, 207, and 208a.
Uses ROUNDOFF and view.
```

例程 setwindow 定义了投影平面上的视见窗口（由中心和维度给出）。

```
196c <setwindow 196c>≡
    void setwindow(Real cu, Real cv, Real su, Real sv)
    {
      if (fabs(su) < ROUNDOFF|| fabs(sv) < ROUNDOFF)
        error("invalid window size");
      view->c.u = cu; view->c.v = cv;
      view->s.u = su; view->s.v = sv;
    }
Defines:
  setwindow, used in chunks 197c, 207, and 208a.
Uses ROUNDOFF and view.
```

例程 setprojection 定义了投影类型。

```
197a <setprojection 197a>≡
    void setprojection(int type)
    {
      if (type ! = PERSPECTIVE && type ! = ORTHOGRAPHIC)
        error("invalid projection type");
      view->type = type;
```

```
    }
Defines:
  setprojection, used in chunks 197c, 207, and 208a.
Uses ORTHOGRAPHIC, PERSPECTIVE, and view.
```

例程 setviewport 定义了设备中所支持的图像区域。

```
197b <setviewport 197b>≡
  void setviewport(Real l, Real b, Real r, Real t, Real n, Real f)
  {
    if(fabs(r-l) < ROUNDOFF|| fabs(t-b) < ROUNDOFF)
      error("invalid viewport");
    view->sc_min.x = l; view->sc_max.x = r;
    view->sc_min.y = b; view->sc_max.y = t;
    view->sc_min.z = n; view->sc_max.z = f;
  }
Defines:
  setviewport, used in chunks 197c and 208b.
Uses ROUNDOFF, view, and viewport 208b.
```

例程 setviewdefaults 定义了默认的视见参数。

```
197c <setviewdefaults 197c>≡
  void setviewdefaults(void)
  {
    setviewpoint(0.0,-5.0,0.0);
    setviewnormal(0.0,1.0,0.0);
    setviewup(0.0,0.0,1.0);
    setviewdistance(1.0);
    setviewdepth(1.0,100000.0);
    setwindow(0.0,0.0,0.41421356,0.31066017);
    setprojection(PERSPECTIVE);
    setviewport(0.,0.,320.,240.,-32768.,32767.);
  }
Defines:
  setviewdefaults, used in chunk 206a.
Uses PERSPECTIVE, setprojection 197a, setviewdepth 196b,
  setviewdistance 196a, setviewnormal 195c, setviewpoint 195b,
  setviewport 197b, setviewup 195d, and setwindow 196c.
```

11.2 视见转换

本节将讨论视见转换，并根据参考空间的定义展示如何计算对应的矩阵。

11.2.1 相机转换

相机转换将三维空间坐标系映射至相机坐标系中。对此，首先将投影中心移至当前原点处，如下所示：

$$A = \begin{pmatrix} I & -V \\ 0 & 1 \end{pmatrix}$$

随后，旋转操作将相机的参考坐标系 $\{u, v, n\}$ 与当前空间的标准基 $\{e_1, e_2, e_3\}$ 对齐，该标准基由向量 $e_1 = (1, 0, 0)$，$e_2 = (0, 1, 0)$ 和 $e_3 = (0, 0, 1)$ 构成。

当计算参考坐标系的向量时，需要使用视见方向 n 和视见的上向量 U。随后将向量 U 投影至相机平面上（同时执行标准化操作）并与 n 正交。

$$v = \frac{U - (U \cdot n)n}{|U - (U \cdot n)n|}$$

接下来将获取与 n 和 v 垂直的单位向量 u，这将使用到叉积运算 $u = n \times v$。需要注意的是，相机系统的方向与场景系统的方向 $\{e_1, e_2, e_3\}$ 相反。这一点可从叉积的计算顺序中看出。

下列正交矩阵表示为一个旋转矩阵，并将坐标系 $\{e_1, e_2, e_3\}$ 映射至坐标系 $\{u, v, n\}$ 中。

$$B = \begin{pmatrix} u^T & 0 \\ v^T & 0 \\ n^T & 0 \\ 0 & 1 \end{pmatrix}$$

基于上述矩阵的连接操作，相机转换由矩阵 $V = BA$ 给出。

$$V = \begin{pmatrix} u_x & u_x & u_x & -u.V_p \\ v_y & v_y & v_y & -v.V_p \\ n_z & n_z & n_z & -n.V_p \\ 0 & 0 & 0 & 1 \end{pmatrix}$$

例程 makeviewV 利用视见参数计算矩阵 V。

```
198 <make view v 198>≡
  void makeviewV(void)
  {
    Vector3 n,u,v,t;

    n = view->normal;
    v = v3_sub(view->up, v3_scale(v3_dot(view->up, n), n));
```

第 11 章 视见相机转换

```
    if (v3_norm(v) < ROUNDOFF)
      error("view up parallel to view normal");
    v = v3_unit(v);
    u = v3_cross(n, v);
    t.x = v3_dot(view->center, u);
    t.y = v3_dot(view->center, v);
    t.z = v3_dot(view->center, n);
    view->V = m4_ident();
    view->V.r1.x =  u.x; view->V.r2.x =  v.x; view->V.r3.x =  n.x;
    view->V.r1.y =  u.y; view->V.r2.y =  v.y; view->V.r3.y =  n.y;
    view->V.r1.z =  u.z; view->V.r2.z =  v.z; view->V.r3.z =  n.z;
    view->V.r1.w = -t.x; view->V.r2.w = -t.y; view->V.r3.w = -t.z;
    makeviewVi();
  }
Defines:
  makeviewV, used in chunks 206, 207a, and 209.
Uses makeviewVi 199, ROUNDOFF, and view.
```

例程 makeviewVi 计算相机的逆转换。

```
199 <make view vi 199>≡
  void makeviewVi(void)
  {
    Vector3 n,u,v,t;
    view->Vinv = m4_ident();
    n = view->normal;
    v = v3_sub(view->up, v3_scale(v3_dot(view->up, n), n));
    if (v3_norm(v) < ROUNDOFF)
      error("view up parallel to view normal");
    v = v3_unit(v);
    u = v3_cross(n, v);
    t = view->center;
    view->Vinv = m4_ident();
    view->Vinv.r1.x = u.x; view->Vinv.r2.x = u.y; view->Vinv.r3.x = u.z;
    view->Vinv.r1.y = v.x; view->Vinv.r2.y = v.y; view->Vinv.r3.y = v.z;
    view->Vinv.r1.z = n.x; view->Vinv.r2.z = n.y; view->Vinv.r3.z = n.z;
    view->Vinv.r1.w = t.x; view->Vinv.r2.w = t.y; view->Vinv.r3.w = t.z;
  }
Defines:
  makeviewVi, used in chunk 198.
Uses ROUNDOFF and view.
```

11.2.2 剪裁转换

剪裁转换将相机的坐标系映射至标准化坐标系中。其中，视见体表示为一个四棱锥，其顶点位于原点处，向量基位于平面 $z=1$ 上。

对此，首先执行剪切操作，以使图像中心与 z 轴对齐。需要注意的是，由于图像窗口无法居中，因而这一步骤不可或缺。

$$D = \begin{pmatrix} 1 & 0 & -c_u/d & 0 \\ 0 & 1 & -c_v/d & 0 \\ 0 & 0 & 1 & 0 \\ 0 & 0 & 0 & 1 \end{pmatrix}$$

随后执行缩放操作，以使视见金字塔的基向量映射至平面 $z=1$ 上的单位正方形中。

$$R = \begin{pmatrix} d/(s_u f) & 0 & 0 & 0 \\ 0 & d/(s_v f) & 0 & 0 \\ 0 & 0 & 1/f & 0 \\ 0 & 0 & 0 & 1 \end{pmatrix}$$

基于上述矩阵的连接操作，剪裁转换由矩阵 $C = RD$ 得到。

$$C = \begin{pmatrix} d/(s_u f) & 0 & -c_u/(s_u f) & 0 \\ 0 & d/(s_v f) & -c_v/(s_u f) & 0 \\ 0 & 0 & 1/f & 0 \\ 0 & 0 & 0 & 1 \end{pmatrix}$$

例程 makeviewC 计算矩阵 C。

```
200 <make view c 200>≡
  void makeviewC(void)
  {
    view->C = m4_ident();
    view->C.r1.x = view->dist / (view->s.u * view->back);
    view->C.r2.y = view->dist / (view->s.v * view->back);
    view->C.r3.z = 1 / view->back;
    view->C.r1.z = - view->c.u / (view->s.u * view->back);
    view->C.r2.z = - view->c.v / (view->s.v * view->back);
    makeviewCi();
  }
Defines:
  makeviewC, used in chunks 206-8.
Uses makeviewCi 201 and view.
```

第 11 章 视见相机转换

例程 makeviewCi 计算矩阵 C 的逆矩阵。

```
201 <make view ci 201>≡
  void makeviewCi(void)
  {
    view->Cinv = m4_ident();
    view->Cinv.r1.x = (view->s.u * view->back) / view->dist;
    view->Cinv.r2.y = (view->s.v * view->back) / view->dist;
    view->Cinv.r3.z = view->back;
    view->Cinv.r1.z = (view->c.u * view->back) / view->dist;
    view->Cinv.r2.z = (view->c.v * view->back) / view->dist;
  }
Defines:
  makeviewCi, used in chunk 200.
Uses view.
```

11.2.3 透视转换

透视转换将下列标准化视见金字塔

$$x = \pm z, \quad y = \pm z, \quad z = z_{\min}, \quad z = 1$$

转换为平行六面体，如下所示。

$$-1 \leqslant x \leqslant 1, \quad -1 \leqslant y \leqslant 1, \quad 0 \leqslant z \leqslant 1$$

其中，$z_{\min} = n/f$。该坐标变换包含了投影转换，透视使用了与直线投影对应的理想点的投影中心。这里首先执行平移操作，并将 $z_{\min}$ 作为投影空间 $\mathbb{RP}^3$ 的原点。

$$E = \begin{pmatrix} 1 & 0 & 0 & 0 \\ 0 & 1 & 0 & 0 \\ 0 & 0 & 1 & -z_{\min} \\ 0 & 0 & 0 & 1 \end{pmatrix}$$

随后执行缩放操作，并将沿 z 方向的区间 $\{z_{\min}, 1\}$ 映射至 $[0, 1]$。

$$F = \begin{pmatrix} 1 & 0 & 0 & 0 \\ 0 & 1 & 0 & 0 \\ 0 & 0 & 1/(1-z_{\min}) & 0 \\ 0 & 0 & 0 & 1 \end{pmatrix}$$

最后执行投影转换，如下所示。

$$G = \begin{pmatrix} 1 & 0 & 0 & 0 \\ 0 & 1 & 0 & 0 \\ 0 & 0 & 1 & 0 \\ 0 & 0 & (1-z_{min})/z_{min} & 1 \end{pmatrix}$$

透视转换通过矩阵 $P = GFE$ 给出，如下所示。

$$P = \begin{pmatrix} 1 & 0 & 0 & 0 \\ 0 & 1 & 0 & 0 \\ 0 & 0 & 1/(1-z_{min}) & -z_{min}/(1-z_{min}) \\ 0 & 0 & 1 & 0 \end{pmatrix}$$

需要注意的是，在执行完投影转换后，还需要利用 w 执行齐次除法。

例程 makeviewP 计算矩阵 P。

```
202a <make view p 202a>≡
  void makeviewP(void)
  {
    view->P = m4_ident();
    view->P.r3.z = view->back / (view->back - view->front);
    view->P.r3.w = -view->front / (view->back - view->front);
    view->P.r4.z = 1;
    view->P.r4.w = 0;
    makeviewPi();
  }
Defines:
  makeviewP, used in chunks 206 and 207.
Uses makeviewPi 202b and view.
```

例程 makeviewPi 计算矩阵 P 的逆矩阵。

```
202b <make view pi 202b>≡
  void makeviewPi(void)
  {
    view->Pinv = m4_ident();
    view->Pinv.r3.z = 0;
    view->Pinv.r4.z = - (view->back - view->front) / view->front;
    view->Pinv.r4.w = view->back / view->front;
    view->Pinv.r3.w = 1;
    view->Pinv.r4.w = 0;
  }
Defines:
  makeviewPi, used in chunk 202a.
Uses view.
```

第 11 章 视见相机转换

除了未涉及投影转换之外，对应于正交投影的转换计算类似于透视转换。这里并不打算展示该计算过程的细节内容。正交投影矩阵 O 如下所示：

$$O = \begin{pmatrix} 1/s_u & 0 & 0 & -c_u/s_u \\ 0 & 1/s_v & 0 & -c_v/s_v \\ 0 & 0 & 1/(f-n) & -n/(f-n) \\ 0 & 0 & 1 & 0 \end{pmatrix}$$

例程 **makeviewO** 计算矩阵 O。

```
203a <make view o 203a>≡
  void makeviewO(void)
  {
    view->C = m4_ident();
    view->C.r1.x = 1 / view->s.u;
    view->C.r2.y = 1 / view->s.v;
    view->C.r3.z = 1 / (view->back - view->front);
    view->C.r1.w = - view->c.u / view->s.u;
    view->C.r2.w = - view->c.v / view->s.v;
    view->C.r3.w = - view->front / (view->back - view->front);
    view->P = m4_ident();
    makeviewOi();
  }
Defines:
  makeviewO, used in chunk 208a.
Uses makeviewOi 203b and view.
```

例程 **makeviewOi** 计算矩阵 O 的逆矩阵。

```
203b <make view oi 203b>≡
  void makeviewOi(void)
  {
    view->Cinv = m4_ident();
    view->Cinv.r1.x = view->s.u;
    view->Cinv.r2.y = view->s.v;
    view->Cinv.r3.z = (view->back - view->front);
    view->Cinv.r1.w = view->c.u;
    view->Cinv.r2.w = view->c.v;
    view->Cinv.r3.w = view->front;
    view->Pinv = m4_ident();
  }
Defines:
  makeviewOi, used in chunk 203a.
Uses view.
```

11.2.4 设备转换

设备转换将标准化可见体映射至图形设备的图像的支持空间中。需要注意的是，此处包含了深度信息，大多数设备均对此予以支持，如 Z 缓冲区。

首先执行平移操作，因而顶点 (-1, -1, 0) 可映射至原点处；随后针对 x 和 y 执行 1/2 缩放操作，这将生成标准化的设备坐标（NDC）空间。

$$K = \begin{pmatrix} 0.5 & 0 & 0 & 0.5 \\ 0 & 0.5 & 0 & 0.5 \\ 0 & 0 & 1 & 0 \\ 0 & 0 & 0 & 1 \end{pmatrix}$$

NDC 空间映射至设备的比例。

$$L = \begin{pmatrix} \Delta_X & 0 & 0 & X_{\min} \\ 0 & \Delta_Y & 0 & Y_{\min} \\ 0 & 0 & \Delta_Z & Z_{\min} \\ 0 & 0 & 0 & 1 \end{pmatrix}$$

其中，$\Delta_i = (i_{\max} - i_{\min})$ 表示为图像三维坐标系的维度。

最后，通过下列矩阵和 floor 函数执行最近整数坐标的舍入操作。

$$M = \begin{pmatrix} 1 & 0 & 0 & 0.5 \\ 0 & 1 & 0 & 0.5 \\ 0 & 0 & 1 & 0.5 \\ 0 & 0 & 0 & 1 \end{pmatrix}, \ \text{floor}(x, y, z)$$

设备转换通过矩阵 $S = KLM$ 给出，如下所示：

$$S = \begin{pmatrix} \Delta_x/2 & 0 & 0 & (\nabla_x+1)/2 \\ 0 & \Delta_y/2 & 0 & (\nabla_y+1)/2 \\ 0 & 0 & \Delta_z & Z_{\min}+0.5 \\ 0 & 0 & 0 & 1 \end{pmatrix}$$

其中，$\Delta_i = (i_{\max} - i_{\min})$，且 $\nabla_i = (i_{\max} + i_{\min})$。

例程 makeviewS 计算矩阵 S。

```
204 <make view s 204>≡
  void makeviewS(void)
  {
    view->S = m4_ident();
```

```
  view->S.r1.x = (view->sc.ur.x - view->sc.ll.x) / 2;
  view->S.r2.y = (view->sc.ur.y - view->sc.ll.y) / 2;
  view->S.r3.z = view->sc.ur.z - view->sc.ll.z;
  view->S.r1.w = (view->sc.ur.x + view->sc.ll.x +1) / 2;
  view->S.r2.w = (view->sc.ur.y + view->sc.ll.y +1) / 2;
  view->S.r3.w = view->sc.ll.z + 0.5;
  makeviewSi();
}
Defines:
  makeviewS, used in chunks 206a and 208b.
Uses makeviewSi 205 and view.
```

例程 makeviewSi 计算矩阵 S 的逆矩阵。

```
205 <make view si 205>≡
  void makeviewSi(void)
  {
    view->Sinv = m4_ident();
    view->Sinv.r1.x = 2 / (view->sc.ur.x - view->sc.ll.x);
    view->Sinv.r2.y = 2 / (view->sc.ur.y - view->sc.ll.y);
    view->Sinv.r3.z = 1 / (view->sc.ur.z - view->sc.ll.z);
    view->Sinv.r1.w = - (view->sc.ur.x + view->sc.ll.x +1) /
                       (view->sc.ur.x - view->sc.ll.x);
    view->Sinv.r2.w = - (view->sc.ur.y + view->sc.ll.y +1) /
                       (view->sc.ur.y - view->sc.ll.y);
    view->Sinv.r3.w = - (view->sc.ll.z + 0.5) /
                       (view->sc.ur.z - view->sc.ll.z);
  }
Defines:
  makeviewSi, used in chunk 204.
Uses view.
```

11.2.5 转换序列

本节所讨论的视见转换将应用于三维场景的对象序列中。该应用取决于对象的几何描述。

如果对象采用参数形式加以描述，则可按照下列顺序采用直接转换，如图 11.2 所示。

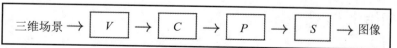

图 11.2　直接转换

如果对象采用隐式形式描述，则按照下列顺序使用逆转换，如图 11.3 所示。

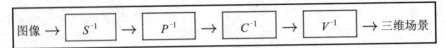

图 11.3　逆转换

针对视见系统的操作组织方式，上述两个转换序列提供了两种不同的策略。第一种策略由直接顺序转换加以定义，对应于以场景为中心的方法，如图 11.4 所示。第二种策略采用逆序转换加以定义，对应于以图像为中心的方法，如图 11.5 所示。

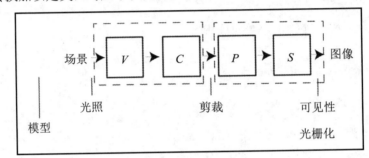

图 11.4　以对象为中心的视见方法

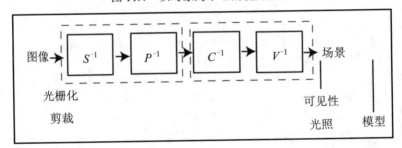

图 11.5　以图像为中心的视见方法

11.3　视见规范

视见转换通过高层例程予以执行，并定义了与每个视见转换相关的参数分组。

11.3.1　初始化

例程 initview 初始化视见参数并计算转换矩阵。

```
206a <initview 206a>≡
  View* initview(void)
  {
    setview(NEWSTRUCT(View));
    setviewdefaults();
    makeviewV();
    makeviewC();
    makeviewP();
    makeviewS();
    return getview();
  }
Defines:
  initview, used in chunk 209.
Uses getview 195a, makeviewC 200, makeviewP 202a, makeviewS 204, makeviewV
198, setview 194c, setviewdefaults 197c, and View 193.
```

11.3.2　相机

例程 lookat 和 camera 定义了相机转换参数。

```
206b <lookat 206b>≡
  void lookat(View *v, Real vx, Real vy, Real vz,
              Real px, Real py, Real pz, Real ux, Real uy, Real uz)
  {
    setview(v);
    setviewpoint(vx, vy, vz);
    setviewnormal(px - vx, py - vy, pz - vz);
    setviewup(ux, uy, uz);
    makeviewV();
  }
Defines:
  lookat, used in chunk 209.
Uses makeviewV 198, setview 194c, setviewnormal 195c, setviewpoint 195b,
setviewup 195d, and View 193.

207a <camera 207a>≡
  void camera(View *v, Real rx, Real ry, Real rz, Real nx, Real ny, Real
  nz, Real ux, Real uy, Real uz, Real deye)
  {
    setview(v);
    setviewup(ux,uy,uz);
    setviewnormal(nx,ny,nz);
```

```
            setviewpoint(rx - (v->normal.x*deye),
                         ry - (v->normal.y*deye),
                         rz - (v->normal.z*deye));
            makeviewV();
         }
Defines:
    camera, never used.
Uses makeviewV 198, setview 194c, setviewnormal 195c, setviewpoint 195b,
setviewup 195d, and View 193.
```

11.3.3 透视

perspective 和 frustum 定义了与剪裁和透视投影转换相关的参数。

```
207b <perspective 207b>≡
    void perspective(View *v, Real fov, Real ar, Real near, Real far)
    {
       setview(v);
       setprojection(PERSPECTIVE);
       setviewdistance(near);
       setviewdepth(near,far);
       if (ar < ROUNDOFF)
          error("illegal aspect ratio");
       setwindow(0, 0, tan(fov/2) * near, (tan(fov/2) * near)/ar);
       makeviewC();
       makeviewP();
    }
Defines:
    perspective, used in chunks 193 and 209.
Uses makeviewC 200, makeviewP 202a, PERSPECTIVE, ROUNDOFF, setprojection
197a, setview 194c, setviewdepth 196b, setviewdistance 196a, setwindow 196c,
and View 193.

207c <frusntrum 207c>≡
    void frustum(View *v, Real l, Real b, Real r, Real t, Real near, Real far)
    {
       setview(v);
       setprojection(PERSPECTIVE);
       setviewdistance(near);
       setviewdepth(near,far);
       setwindow((l+r)/2, (b+t)/2, (r-l)/2, (t-b)/2);
       makeviewC();
```

第 11 章 视见相机转换

```
    makeviewP();
  }
Defines:
  frustum, never used.
Uses makeviewC 200, makeviewP 202a, PERSPECTIVE, setprojection 197a,
setview 194c, setviewdepth 196b, setviewdistance 196a, setwindow 196c, and
View 193.
```

例程 orthographic 定义了与剪裁和正交投影转换相关的参数。

```
208a <orthographic 208a>≡
  void orthographic(View *v, Real l, Real b, Real r, Real t, Real near,
                    Real far)
  {
    setview(v);
    setprojection(ORTHOGRAPHIC);
    setviewdistance(near);
    setviewdepth(near,far);
    setwindow((l+r)/2, (b+t)/2, (r-l)/2, (t-b)/2);
    makeviewC();
    makeviewO();
  }
Defines:
  orthographic, never used.
Uses makeviewC 200, makeviewO 203a, ORTHOGRAPHIC, setprojection 197a,
setview 194c, setviewdepth 196b, setviewdistance 196a, setwindow 196c, and
View 193.
```

11.3.4 设备

例程 viewport 定义了与设备转换相关的参数。

```
208b <viewport 208b>≡
  void viewport(View *v, Real l, Real b, Real w, Real h)
  {
    setview(v);
    setviewport(l,b,l+w,b+h,-32767.,32767.);
    makeviewS();
  }
Defines:
  viewport, used in chunks 197b and 209.
Uses makeviewS 204, setview 194c, setviewport 197b, and View 193.
```

11.3.5 定义三维场景描述语言中的视见机制

针对三维场景描述语言中的视见机制,例程 view_parse 实现了相关命令。

```
209 <parse view 209>≡
  Val view_parse(int pass, Pval *pl)
  {
    Val v;
    if (pass == T_EXEC) {
      View *view = initview();
      Vector3 ref = pvl_get_v3(pl, "from", v3_make(0,-5,0));
      Vector3 at = pvl_get_v3(pl, "at", v3_make(0,0,0));
      Vector3 up = pvl_get_v3(pl, p", v3_make(0,0,1));
      double fov = pvl_get_num(pl, "fov", 90);
      double w = pvl_get_num(pl, "imgw", 320);
      double h = pvl_get_num(pl, "imgh", 240);
      lookat(view, ref.x, ref.y, ref.z, at.x, at.y, at.z, up.x, up.y, up.z);
      setviewpoint(ref.x, ref.y, ref.z);
      setviewnormal(at.x - ref.x, at.y - ref.y, at.z - ref.z);
      makeviewV();
      perspective(view, fov * DTOR, w/h, 1.0, 100000.0);
      viewport(view, 0.,0., w, h);
      v.type = CAMERA;
      v.u.v = view;
    }
    return v;
  }
Defines:
 view_parse, never used.
Uses initview 206a, lookat 206b, makeviewV 198, perspective 207b,
setviewnormal 195c, setviewpoint 195b, View 193, view, and viewport 208b.
```

11.4 补充材料

本章通过虚拟相机讨论了视见转换及其规范。
VIEW 库中的 API 包含以下例程:

```
View* initview(void);
void lookat(View *v,Real vx, Real vy, Real vz, Real px, Real py, Real
```

```
                pz, Real ux, Real uy, Real uz);
void camera(View *v, Real rx, Real ry, Real rz,
            Real nx, Real ny, Real nz, Real ux, Real uy, Real uz, Real deye);
void perspective(View *v, Real fov, Real ar, Real near, Real far);
void orthographic(View *v, Real l, Real b, Real r, Real t, Real near,
                  Real far);
void frustum(View *v, Real l, Real b, Real r, Real t, Real near, Real far);
void viewport(View *v, Real l, Real b, Real w, Real h);

void setview(View *v);
View *getview(void);
void setviewpoint(Real x, Real y, Real z);
void setviewnormal(Real x, Real y, Real z);
void setviewup(Real x, Real y, Real z);
void setviewdistance(Real d);
void setviewdepth(Real front, Real back);
void setwindow(Real cu, Real cv, Real su, Real sv);
void setprojection(int type);
void setviewport(Real l, Real b, Real r, Real t, Real n, Real f);
void setviewdefaults(void);

void makeviewV(void);
void makeviewC(void);
void makeviewO(void);
void makeviewP(void);
void makeviewS(void);
void makeviewVi(void);
void makeviewCi(void);
void makeviewOi(void);
void makeviewPi(void);
void makeviewSi(void);
```

11.5 本章练习

利用 VEIW 库编写一个程序，并绘制三维场景的线框图。其中，程序的输入内容表示为场景描述语言，同时应确定虚拟相机和一个或多个对象。

第 12 章 视见的表面剪裁

剪裁操作是计算机图形学中的基本几何工具之一，并在几何建模处理和视见问题中饰演了重要的角色。本章将讨论事件处理过程中的表面剪裁操作。

12.1 剪裁操作的基本知识

本节所讨论的操作用于确定包含于环境空间剪裁区域内的、图形对象的点集。

12.1.1 空间剪裁

当给定维度 $n-1$ 的一个封闭和连接的表面 $M \subset \mathbb{R}^n$，M 将空间划分为两个区域 A 和 B。其中，M 表示为公共边界。$\mathbb{R}^n$ 的子集 S 的剪裁操作涉及子集 $S \cap A$ 和 $S \cap B$ 的确定。这里，表面 M 称作剪裁表面。

需要注意的是，对于剪裁操作来说，可分性是表面 M 的一个基本属性，进而将环境空间的划分为两个区域。

总体而言，剪裁操作涉及 3 个不同问题的处理过程：
- ❑ 相交。确定边界处（由 M 给出）S 中的点。
- ❑ 分类。对包含于区域 A 和 B 中的 S 的点进行分类。
- ❑ 结构化。利用上述结果构建集合 $S \cap A$ 和 $S \cap B$。

剪裁计算取决于子集 S 的维度及其几何表达。在 $\mathbb{R}^3$ 中，S 可以是一个点、一个曲面、一个表面或者是一个实体。S 的表达可通过图元、构建模式或分解加以确定。

另一个影响剪裁操作复杂度的因素是剪裁表面的几何体 M。特别地，如果 M 为凸体，则问题可得到简化。

12.1.2 剪裁和视见

在视见处理过程中，剪裁操作的目的是确定哪些对象处于虚拟相机视域范围内，此类对象将被处理，而其他对象则被丢弃。视见处理过程中的剪裁操作包含两个原因：首先，剪裁可防止视见相反方向上的点被错误地投影至图像平面上；其次，剪裁可提升工作效率，也就是说，不会对未出现于图像中的对象进行处理。

在第 11 章中曾讨论到，剪裁视见体称作视见平截头体，对应于截断后的金字塔。相应地，应通过某种方式谨慎地选取标准化剪裁空间，进而简化当前问题的处理方案。据此，视见平截头体由半空间的相交结果确定。

$$-x \leqslant z \leqslant x,\ -y \leqslant z \leqslant z,\ z_{min} \leqslant z \leqslant 1 \tag{12.1}$$

该选择方案极大地简化了相交结果的计算过程，而且，由于视见平截头体为凸体，因而分类操作简化为确定相对于式（12.1）中 6 个半空间的点位置。

另一个重要的观察结果是，当子集 S 完全包含于区域 A 或 B 中，剪裁问题可方便地予以解决。这一情况完全可通过点分类加以确定，且不需要进行相交计算或结构化操作。

上述计算策略可在剪裁算法中结合使用。首先，通过适宜的坐标系，问题简化为一种标准情形。其次，检测包含快速解决方案的简单情形。最后，即可解决较为复杂的问题。

12.2 剪裁简单情形

当子集 S 完全包含于剪裁表面 M 所分隔的某个区域中，即会出现剪裁中的简单情形。对此，存在以下两种情况：S 位于视见体的外部，因而可从当前处理过程中移除；S 位于视见体内部，并可投影至虚拟屏幕上。

12.2.1 简单拒绝

当检测某个多边形是否完全包含于剪裁体的外部区域时，可相对于视见体表面的支撑平面确定的半空间执行多边形顶点分类操作。这一类平面将空间划分为 27 个区域。针对于此，每一个顶点可定义一个 6 位代码，表示该顶点所在的区域。该代码表示当前区域和某个半空间之间的关系，例如，1 表示当前区域未处于剪裁金字塔的同一半空间内。

例程 clipcode 计算顶点代码。

```
213a <clipcode 213a>≡
  int clipcode(Real h, Vector3 v)
  {
    int c = 0;
    if (v.y >  v.z) c|= 01;
    if (v.y < -v.z) c|= 02;
    if (v.x >  v.z) c|= 04;
    if (v.x < -v.z) c|= 010;
    if (v.z >  1  ) c|= 020;
```

第 12 章 视见的表面剪裁

```
    if (v.z < h ) c| = 040;
    return(c);
  }
Defines:
  clipcode, used in chunk 213.
```

例程 cull_poly3 确定三角形是否位于剪裁区域的外部区域内,并针对各顶点采用了分类操作。如果全部顶点均位于视见体外部半空间中,则该三角形将被拒绝。

```
213b <cullpoly3 213b>≡
  int cull_poly3(Real h, Poly *p)
  {
    int i, c[3] ;
    for (i = 0; i < p->n; i++)
      c[i] = clipcode(h, p->v[i] );
    return (c[0] & c[1] & c[2] );
  }
Defines:
  cull_poly3, used in chunk 215.
Uses clipcode 213a.
```

12.2.2 简单接受

例程 inside_frustum 确定某个点是否位于视见体内部,该例程用于执行凸体对象的简单接受测试,如三角形或复杂对象的包围盒。

```
213c <inside frustum 213c>≡
  int inside_frustum(Real h, Vector3 v)
  {
    return (clipcode(h, v) == 0);
  }
Defines:
  inside_frustum, never used.
Uses clipcode 213a.
```

12.2.3 包含相反方向的面元

在实体对象中,与视见方向相反的边界面元处于不可见状态,并应予以移除。

在边界表面中,包含相反方向的面元处于可见状态。在计算光照时,一般需要考虑到这一点。

例程 is_backfacing 用于确定多边形方向是否与视见方向相反。这一计算过程涉及多边形法线和视见方向间的内积，如图 12.1 所示。

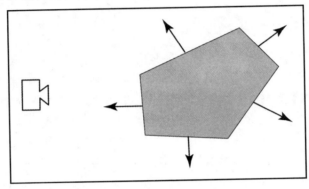

图 12.1　相机与包含法线的对象

```
214 <is backfacing 214>≡
  int is_backfacing(Poly *p, Vector3 v)
  {
    Vector3 n = poly_normal(p);
    return (v3_dot(n, v) < 0)? TRUE : FALSE;
  }
Defines:
  is_backfacing, never used.
```

12.3　两步剪裁

实现多边形剪裁的简单方法分为两个步骤。在第一个步骤中，将执行三维剪裁操作；在第二个步骤中，则实现二维剪裁操作。该问题可通过下列方式加以组织：

- 三维部分。剪裁与视见体正面相关的多边形。该操作需要在投影转换之前在三维空间内予以执行，进而避免将位于相机后方的顶点投影至图像平面上。
- 二维部分。相对于图像矩形，剪裁已投影的多边形。该操作在光栅化之前通过解析方式执行，或者在光栅化过程中以近似方式执行。

因此，远平面剪裁算法包含以下步骤：

（1）简单情形分析。
（2）基于正面的剪裁。
（3）基于图像矩形的二维剪裁。

第 12 章 视见的表面剪裁

例程 hither_clip 实现步骤（1）和（2），且例程 render 执行步骤（3）。

```
215 <hither clip 215>≡
  int hither_clip(Real h, Poly *p, void (*render)(), void (*plfree)())
  {
    if (cull_poly3(h, p)) {
      plfree(p); return FALSE;
    } else {
      return hclip(h, p, render, plfree);
    }
  }
Defines:
  hither_clip, never used.
Uses cull_poly3 213b and hclip 216a.
```

其中，步骤（1）（简单情形分析）为可选项。这一预处理过程旨在提升算法效率，并通过例程 cull_poly3 予以实现。通过这种方式，位于剪裁体外部的所有多边形均被剔除。

这里所描述的方法具有某些重要特征。例程 hclip 实现了递归算法，并通过三角形的二叉子划分执行分析式剪裁。该算法执行三角形列表剪裁，其中，列表中的第一个三角形包含了几何和属性信息。

算法的一般结构可描述为：首先，三角形顶点相对于正面进行分类，这可通过平面正空间中的顶点数量加以确定，其中包含以下 3 种情况：

❏ $n = 0$。全部三角形位于平面的负空间，并可简单地对其拒绝。
❏ $n = 3$。全部三角形位于平面的正空间，并被可视化（简单接受）。
❏ $n = 1$ 或 $n = 2$。三角形与平面相交，需要执行剪裁操作。

例程 classify_vert 计算 n。

```
216a <hclip 216a>≡
  int hclip(Real h, Poly *p, void (*render)(), void (*plfree)())
  {
    Poly *pa, *pb, *a, *b;
    int n, i0, i1, i2;
    double t;

    switch (classify_vert(p, h)) {
    case 0: plfree(p); return FALSE;
    case 3: render(p); return TRUE;
    case 2: case 1:
      if (EDGE_CROSS_Z(p, 0, 1, h)) {
        i0 = 0; i1 = 1; i2 = 2;
```

```
      } else if (EDGE_CROSS_Z(p, 1, 2, h)) {
         i0 = 1; i1 = 2; i2 = 0;
      } else if (EDGE_CROSS_Z(p, 2, 0, h)) {
         i0 = 2; i1 = 0; i2 = 1;
      }
   }
   t = (p->v[i1].z - h) / (p->v[i1].z - p->v[i0].z);

   a = pa = plist_alloc(n = plist_lenght(p), 3);
   b = pb = plist_alloc(n, 3);
   while (p != NULL) {
      poly_split(p, i0, i1, i2, t, pa, pb);
      p = p->next; pa = pa->next; pb = pb->next;
   }
   return hclip(h, a, render, plfree)| hclip(h, b, render, plfree);
}
Defines:
 hclip, used in chunk 215.
Uses classify_vert 218, EDGE_CROSS_Z 216b, and poly_split 217.
```

在最后一种情况中，三角形被划分为两个三角形，每个三角形都通过递归方式提交至当前算法中。

当执行细分时，将选取与正面相交的一条边。该边将在平面交点处进行细分，如图12.2所示。

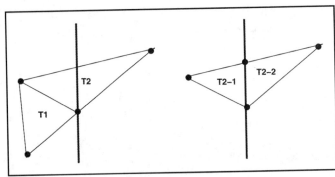

图12.2 平面细分

宏 EDGE_CROSS_Z 确定边 (v_0, v_1) 是否穿越平面 $z = h$。

```
216b <edge cross z 216b>≡
   #define EDGE_CROSS_Z(P, V0, V1, H) \
          ((REL_GT((P)->v[V0].z, H) && REL_LT((P)->v[V1].z, H)) \
```

```
            || (REL_LT((P)->v[V0].z, H) && REL_GT((P)->v[V1].z, H)))
Defines:
  EDGE_CROSS_Z, used in chunk 216a.
```

例程 poly_split 对三角形进行细分。

```
217 <poly split 217>≡
  void poly_split(Poly *p, int i0, int i1, int i2, Real t, Poly *pa,
                  Poly *pb)
  {
    Vector3 vm = v3_add(p->v[i1], v3_scale(t, v3_sub(p->v[i0],
    p->v[i1] )));
    pa->v[0] = p->v[i1] ; pa->v[1] = p->v[i2] ;
    pa->v[2] = pb->v[0] = vm;
    pb->v[1] = p->v[i2] ; pb->v[2] = p->v[i0] ;
  }
Defines:
  poly_split, used in chunk 216a.
```

在最后一种情况下，当细分产生的两个三角形被算法递归处理时，一个三角形将被拒绝，另一个三角形将被最后一次细分，如图 12.3 所示。

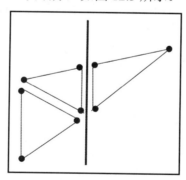

图 12.3　平面递归细分

最后，在第二种情况下被接受的多边形将被移至方法的最后一部分中，并相对于图像矩形在二维空间内被剪取。

例程 classify_vert 计算平面 $z = h$ 正空间中顶点的数量。

```
218 <classify vert 218>≡
  static int classify_vert(Poly *p, Real h)
  {
    int i, k = 0;
```

```
    for (i = 0; i < 3; i++)
      if (REL_GT(p->v[i].z, h)) k++;
    for (i = 0; i < 3; i++)
      if (k > 0 && REL_EQ(p->v[i].z, h)) k++;
    return k;
  }
Defines:
  classify_vert, used in chunk 216a.
```

在相对于平面 $z = h$ 的顶点分类过程中，应留意一种特殊情形。其中，递归算法生成了非常规位置的顶点，其原因在于，边细分生成了恰好位于剪裁面上的顶点。在算法递归中，顶点分类应考虑到何种情况。对此，可通过例程 classify_vert 和宏 EDGE_CROSS_Z 予以处理，进而避免模型的几何形状和拓扑结构间的不一致性。

12.4 序列剪裁

12.3 节描述的基于细分算法的剪裁操作基于二分三角形简化了结构化问题。本节针对 n 边通用多边形介绍一种算法，并相对于视见体在三维环境中执行整体剪裁操作。

该算法所采用的策略可描述为：针对视见体表面确定的每个半空间以序列方式剪裁多边形。在每个步骤中，位于剪裁面正空间中的多边形将被计算，并传递至下一个阶段中。鉴于剪裁体表示为凸体，因而在处理结束时，最终的多边形表示为原多边形与剪裁空间的相交结果。

稍作修改，该算法可用于剪裁任意凸体。

该剪裁算法使用下列代码识别视见体平面。

```
#define LFT 0
#define RGT 1
#define TOP 2
#define BOT 3
#define BAK 4
#define FRT 5
```

例程 poly_clip 针对多边形列表执行 Sutherland-Hodgman 剪裁算法[Sutherland and Hodgman 74]。其中，列表中的第一个多边形包含了几何信息（如顶点位置）；其他多边形则包含了属性值（如颜色、纹理坐标、法线等）。该例程首先基于几何信息计算与剪裁面间的距离，随后根据此类信息针对几何形状和属性执行剪裁操作。其间，距离信息存储于内部数组 dd 中。通过这种方式，poly_clip 首先调用例程 pclip_store 计算剪裁操作

第 12 章 视见的表面剪裁

序列,并于随后调用例程将操作结果转换为输出多边形。如果多边形列表涵盖其他属性的多边形,剪裁算法通过 pclip_apply 和 clip_copy 例程还将应用于多边形列表的尾部。

```
219 <poly clip 219>≡
  int poly_clip(Real h, Poly *p, void (*dispose)(), int chain)
  {
    double dd[MAXD] ;
    Poly *a;
    int n;

    if (p->n > MAXV)
      error("(poly clip) too many vertices");

    n = clip_copy(pclip_store, p, h, dispose, dd);
    if (! chain)
      return n;
    for (a = p->next; a ! = NULL; a = a->next)
      n = clip_copy(pclip_apply, a, h, dispose, dd);

    return n;
  }
Defines:
  poly_clip, never used.
Uses clip_copy 222b, MAXD 222a, MAXV 222a, pclip_apply 221c, and pclip_
  store 220a.
```

例程 pclip_store 通过一个剪裁面执行多边形剪裁。另外,该例程还将执行分类、相交和结构化操作,进而处理输入多边形的各边。当前边由 k0 和 k1 给出,分别表示为首、尾顶点的索引。

顶点根据与剪裁面间的有符号距离进行分类,并存储于数组 dd 中。如果边与平面相交,则计算交点并插入输出多边形的顶点列表中。如果边的尾顶点位于剪裁面的正空间,还需要将其插入输出多边形的顶点列表中。需要注意的是,输入多边形的第一个顶点将于最后处理,其原因在于,对于闭环来说,最后一条边对应于 (v_{n-1}, v_0)。

此外,例程 pclip 执行剪裁的序列操作,增加了剪裁平面代码,并将输入缓冲区与输出缓冲区交换。

```
220a <pclip store 220a>≡
  int pclip_store(int plane, Poly *s, Poly *d, Real h, double *dd)
  {
    int i, k0, k1;
    double d0, d1;
```

```
        for (d->n = k1 = i = 0; i <= s->n ; i++, k1 = (i == s->n)? 0 : i) {
          d1 = plane_point_dist(plane, s->v[k1], h);
          DA(dd, i, plane) = d1;
          if (i ! = 0) {
            if (PLANE_CROSS(d0, d1))
              d->v[d->n++] = v3_add(s->v[k1], v3_scale(d1/(d1-d0),
                 v3_sub(s->v[k0], s->v[k1] )));
            if (ON_POSITIVE_SIDE(d1))
              d->v[d->n++] = s->v[k1] ;
          }
          d0 = d1;
          k0 = k1;
        }
        return (plane++ == FRT)? d->n : pclip_store(plane, d, s, h, dd);
      }
Defines:
 pclip_store, used in chunk 219.
Uses DA 222a, FRT, ON_POSITIVE_SIDE, PLANE_CROSS 221a, and plane_point_
dist 220b.
```

例程 plane_point_dist 计算顶点与平面间的有符号距离。

```
220b <plane point dist 220b>≡
  Real plane_point_dist(int plane, Vector3 v, Real h)
  {
    switch (plane) {
    case LFT: return v.z + v.x;
    case RGT: return v.z - v.x;
    case TOP: return v.z + v.y;
    case BOT: return v.z - v.y;
    case BAK: return 1 - v.z;
    case FRT: return -h + v.z;
    }
  }
Defines:
 plane_point_dist, used in chunk 220a.
Uses BAK, BOT, FRT, LFT, RGT, and TOP.
```

宏 PLANE_CROSS 检测某一条边是否与剪裁面相交，该计算过程基于边顶点与平面间的有符号距离。

```
221a <plane cross 221a>≡
  #define PLANE_CROSS(D0, D1) ((D0) * (D1) < 0)
```

第 12 章 视见的表面剪裁

Defines:
  PLANE_CROSS, used in chunks 220a and 221c.

根据有符号距离，宏 ON_POS_SIDE 检测顶点是否位于剪裁面的正空间内。

221b <on postive side 221b>≡
  #define ON_POS_SIDE(D1) ((D1) >= 0)
Defines:
  ON_POS_SIDE, never used.

例程 pclip_apply 使用与剪裁平面间的距离（通过例程 pclip_store 计算得到，并位于数组 dd 中），并将实际的剪裁操作应用于属性多边形中。

通过这种方式，在不需要显示几何信息的情况下，可以在后续步骤中对具有属性值的多边形重复执行几何裁剪计算。

221c <pclip apply 221c>≡
```
  int pclip_apply(int plane, Poly *s, Poly *d, Real h, double *dd)
  {
    int i, k0, k1;
    double d0, d1;

    for (d->n = k1 = i = 0; i <= s->n ; i++, k1 = (i == s->n)? 0 : i) {
      d1 = DA(dd, i, plane);
      if (i != 0) {
        if (PLANE_CROSS(d0, d1))
          d->v[d->n++] = v3_add(s->v[k1], v3_scale(d1/(d1-d0),
            v3_sub(s->v[k0], s->v[k1] )));
        if (ON_POSITIVE_SIDE(d1))
          d->v[d->n++] = s->v[k1] ;
      }
      d0 = d1;
      k0 = k1;
    }
    return (plane++ == FRT)? d->n : pclip_apply(plane, d, s, h, dd);
  }
```
Defines:
  pclip_apply, used in chunk 219.
Uses DA 222a, FRT, ON_POSITIVE_SIDE, and PLANE_CROSS 221a.

宏 DA、MAXV 和 MAXD 则用于控制对平面距离数组 dd 的访问。

222a <da 222a>≡
  #define MAXV 16

```
  #define MAXD ((MAXV+1)*(FRT+1))

  #define DA(dd, i, p) dd[(p*MAXV)+i]
Defines:
  DA, used in chunks 220a and 221c.
  MAXD, used in chunk 219.
  MAXV, used in chunks 219 and 222b.
Uses FRT.
```

例程 clip_copy 负责算法的内存管理,并分配了两个内部缓冲区,在每个剪裁处理阶段交替用作输入和输出缓冲区。除此之外,该例程还用于确定是否有必要针对剪裁后的多边形分配新的内存空间,并将操作结果复制至相应的数据结构中。

```
222b <clip copy 222b>≡
  int clip_copy(int (*clip_do)(), Poly *p, Real h, void (*dispose)(),
                  double *dd)
  {
    Vector3 vs[MAXV], vd[MAXV] ;
    Poly s = {NULL, 0, vs}, d = {NULL, 0, vd};

    poly_copy(p, &s);
    if (clip_do(0, &s, &d, h, dd) > p->n) {
      if (dispose ! = NULL)
        dispose(p->v);
      p->v = NEWARRAY(s.n, Vector3);
    }
    poly_copy(&s, p);
    return s.n;
  }
Defines:
  clip_copy, used in chunk 219.
Uses MAXV 222a.
```

12.5 补充材料

本章针对视见机制讨论了表面剪裁问题,并对此介绍了相关算法。图 12.4 显示了一个执行剪裁操作并绘制正交投影的程序示例。

第 12 章 视见的表面剪裁

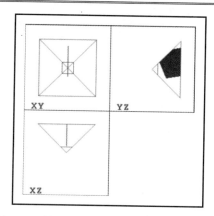

图 12.4　针对视见金字塔（视见体）的剪裁

CLIP 库中的 API 由以下例程构成：

```
int is_backfacing(Poly *p);

int clip(Poly *s);

int clip_sub(Poly *p, void (*render)());
```

12.6　本章练习

（1）编写一个程序，利用细分算法并针对正面（远平面）剪裁三角形，最终结果可利用正交于 z 轴的投影予以可视化。

（2）编写一个程序，利用 Sutherland-Hodgman 算法，并针对视见金字塔（视见体）剪裁四边形，最终结果可利用正交于 z 轴的投影予以可视化。

（3）对练习（2）稍作修改并绘制投影平面。

第 13 章 光 栅 化

视见处理的结果是一幅图像。在三维场景视见环境中，相机视域内的对象投影至图像平面上，并离散化为图像元素（或像素）。本章将讨论执行此类转换的光栅化操作。

13.1 光栅化基础知识

考查图形对象 $\mathcal{O}=(S,f)$，其中包含了几何支撑 $S \subset R^n$、属性函数 $f: S \to R^k$，以及 $\mathbb{R}^n$ 的网格 Δ（分辨率为 $\Delta_1 \times \ldots \times \Delta_n$）。光栅化将确定基于 Δ 的、$\mathcal{O}$ 的表达矩阵。

相应地，较为常见的情形则是二维光栅化，其中，$n=2$ 且 S 表示为平面的二维子集。此时，表达矩阵对应于数字图像，如图 13.1 所示。

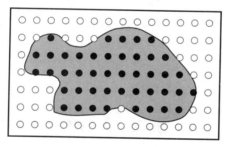

图 13.1 二维光栅化

光栅化处理涉及与支撑 S 关联的、网格 Δ 中单元（C_i）的枚举结果，以及每个单元 C_i 中的属性函数 f 的采样机制。

实际上，光栅化操作将确定网格 Δ 的单元 C_i 是否是 $\mathcal{O}$ 矩阵表达的单元序列中的一部分内容，并可根据单元与集合 S 间的交集予以判断。需要注意的是，该处理过程与第 12 章所讨论的剪裁机制密切相关。

在三维场景视见环境中，对象投影至图像平面上，并于随后执行光栅化操作，取决于 S 的维度，S 的几何支撑可以是一条曲线或者是平面区域。

属性函数提供了图像对象可视化表达的颜色值，因而也称作着色函数。相应地，着色函数将生成三维场景的光照环境。光照计算以及着色函数的采样将在后续章节中加以讨论。

考虑到可见性问题,光栅化和三维视见处理之间的关系较为微妙。理想状态下,可仅对虚拟相机可见的表面执行光栅化操作。在实际操作过程中,可能会存在以下 3 种情况:

- 可见性计算之后的光栅化。
- 可见性计算之前的光栅化。
- 同时计算光栅化和可见性。

第 14 章将处理可见性问题,以及上述 3 种策略的解决方案。本章将讨论光栅化算法,且暂时忽略可见性问题。

13.2 光栅化方法的分类

从本质上讲,依据网格 Δ 光栅化对应于图形对象 O 的几何支撑的离散化结果。我们可根据两种标准对光栅化方法进行分类,即离散化策略和几何描述类型。

离散化策略与如何执行离散化相关,可采用渐增方法或细分方法。其中,渐增式离散化针对几何体的离散化采用了迭代处理;而细分离散化则采用了递归处理。相比之下,渐增式方法更加高效;而细分方法则具有自适应性。

取决于光栅化所依赖的空间,几何描述类型分为内在型和外在型。其中,内在型光栅化基于图形对象的离散化结果;而外在型光栅化则基于空间的离散化结果。相应地,内在型方法在对象的参数空间内进行操作;外在型方法则直接在世界空间内进行操作。

下面将讨论基于上述分类的光栅化方法,并针对多边形区域进一步考查渐增式方法、内在型和外在型几何描述。除此之外,还将针对直线段阐述细分方法、内在型和外在型方法。

13.3 渐增式方法

本节将针对平面区域光栅化介绍渐增式方法。渐增式光栅化算法其构建过程始于两个基本操作,即初始化和遍历操作。其中,初始化操作指定了初始单元,并计算渐增式数据。遍历操作则在当前单元和下一个单元间移动,直至枚举了所有单元。

13.3.1 内在型渐增式光栅化

这里将重点讨论三角形的光栅化,对于图像合成来说,这也是最为重要的例子。而且,任意多边形区域均可实现三角剖分,进而转化为当前示例。

第 13 章 光 栅 化

例程 scan_poly 光栅化凸多边形，并将多边形分解为三角形，随后通过例程 scan_spoly3 进行光栅化。其间，三角剖分过程可描述为：将多边形的第一个顶点与该多边形的其他顶点链接进而形成三角形(v_0, v_{k-1}, v_k)，$k = 2, \ldots n-1$。

```
227a <scan poly 227a>≡
  void scan_poly(Poly *p, Paint *pfun, void *pdata)
  {
    int k;
    for (k = 2; k < p->n; k++)
      scan_spoly3(p, k, pfun, pdata);
  }
Defines:
  scan_poly, never used.
Uses scan_spoly3 227b.
```

三角形格式化将考查以下事实：凸多边形区域与网格水平带的交点由一对边(e_l, e_r)决定。针对 j，存在一个单元序列 $c_{i,j}$，且 $x_l \leqslant i \leqslant x_r$，其中，$(x_l, j) \in e_l$，且 $(x_r, j) \in e_r$。这里，区间 $[x_l, e_r]$（也称作跨度）始于三角形边的参数描述，并采用渐增方式予以计算。

例程 scan_spoly3 实现了三角形的光栅化操作。对应算法由内嵌于两个级别的渐增式处理过程构成，即全局级别（多边形）和局部级别（边）。

在全局级别上，初始化过程将确定初始边对。遍历操作则枚举每个跨度的单元，同时遍历多边形的各条边。在局部级别上，初始化操作计算边的初始位置和渐增值，遍历过程则沿边的方向更新当前位置。

```
227b <scan poly tri 227b>≡
  void scan_spoly3(Poly *p, int n, Paint *pfun, void *pdata)
  {
    Edge *lft, *rgt, *seg, l, r, s;
    int lv, rv, tv = first_vertex(p->v, n);

    rgt = edge(p->v[tv], p->v[rv = NEXT(tv,n)], 'y', &r);
    lft = edge(p->v[tv], p->v[lv = PREV(tv,n)], 'y', &l);
    while (lft && rgt) {
      seg = edge(lft->p, rgt->p, 'x', &s);
      do
        pfun(seg->p, n, lv, lft->t, rv, rgt->t, seg->t, pdata);
      while (seg = increment(seg));
      if (! (rgt = increment(rgt))) { tv = rv;
        rgt = increment(edge(p->v[tv], p->v[rv = NEXT(rv,n)], 'y', &r));
      }
```

```
      if (! (lft = increment(lft))) { tv = lv;
        lft = increment(edge(p->v[tv], p->v[lv = PREV(lv,n)], 'y', &l));
      }
    }
  }
Defines:
  scan_spoly3, used in chunk 227a.
Uses Edge 228b, edge 229a, first_vertex 228a, increment 229b, NEXT, and PREV.
```

例程 first_vertex 计算最低序的顶点，进而确定栅格化的初始边对。

```
228a <first vertex 228a>≡
  static int first_vertex(Vector3 v[], int n)
  {
    if (v[0].y < v[n-1].y)
      return ((v[0].y < v[n].y)? 0: n);
    else
      return ((v[n-1].y < v[n].y)? n-1: n);
  }
Defines:
  first_vertex, used in chunk 227b.
```

宏 PREV 和 NEXT 用于遍历正、负方向中的各条边。

```
#define PREV(K,N)  ((K == N)? 0 : ((K == 0)? N - 1 : N))
#define NEXT(K,N)  ((K == 0)? N : ((K == N)? N - 1 : 0))
```

数据结构 Edge 用于描述一条边，除了增量 n 之外，其中还包含了当前位置 p、位置 i 的增量、当前参数坐标 t，以及参数增量 d。

```
228b <edge structure 228b>≡
  typedef struct Edge {
    int n;
    Vector3 p, i;
    double t, d;
  } Edge;
Defines:
  Edge, used in chunks 227b and 229.
```

例程 edge 利用 x 或 y 方向上的增量生成一条边 (p_0, p_1)。注意，y 方向上的增量执行垂直扫描，这将遍历多边形区域的边界。另外一方面，x 方向上的增量执行水平扫描，这将枚举某个跨度上的单元。

```
229a <edge 229a>≡
  static Edge *edge(Vector3 p0, Vector3 p1, char c, Edge *e)
```

第13章 光栅化

```
  {
    SNAP_XY(p0); SNAP_XY(p1);
    switch (c) {
    case 'x': e->n = ABS(p1.x - p0.x); break;
    case 'y': e->n = p1.y - p0.y; break;
    }
    e->t = 0;
    e->p = p0;
    e->d = (e->n < 1) ? 0.0 : 1.0 / e->n;
    e->i = v3_scale(e->d, v3_sub(p1, p0));
    return e;
  }
Defines:
  edge, used in chunk 227b.
Uses Edge 228b and SNAP_XY.
```

宏 SNAP_XY 将坐标 (x,y) 限制为整型网格 $\mathbb{Z} \times \mathbb{Z}$ 的点。

```
#define SNAP_XY(V) {V.x = rint(V.x), V.y = rint(V.y); }
```

例程 increment 更新边的当前位置和参数，同时还将测试中止条件。

```
229b <increment 229b>≡
  static Edge *increment(Edge *e)
  {
    e->p = v3_add(e->p, e->i);
    e->t += e->d;
    return (e->n-- > 0) ? e : (Edge *)0;
  }
Defines:
  increment, used in chunk 227b.
Uses Edge 228b.
```

每一个由当前算法枚举的单元 $C_{i,j}$ 将传递至绘制例程 pfun 中，进而处理对应的像素。其中，变量 pdata 存储该处理过程中的数据。总体而言，我们已经得到了可与多边形顶点关联的图形对象的属性。例程 pfun 通过双线性插值计算单元中的属性值。

例程 seg_bilerp 实现了双线性插值计算。

```
230a <seg bilerp 230a>≡
  Vector3 seg_bilerp(Poly *p, int n, Real t, int lv, Real lt, int rv,
                     Real rt)
  {
    return v3_bilerp(t, lt, p->v[NEXT(lv,n)], p->v[lv],
                         rt, p->v[PREV(rv,n)], p->v[rv] );
```

```
}
Defines:
  seg_bilerp, never used.
Uses NEXT and PREV.
```

13.3.2 外在型渐增式光栅化

外在型渐增式光栅化扫描图像平面上整数网格的所有单元。扫描过程通过图像矩阵的行和列执行。换而言之，单元 $C_{i,j}$ 通过 $i=1,\ldots,n$，$j=1,\ldots,m$ 予以访问。如果 $(C_{i,j} \cap S) = \phi$，则单元 $C_{i,j}$ 为图形对象 $\mathcal{O}=(S,f)$ 表达中的一部分内容。当 S 表示为平面区域时，则可通过近似方式执行。也就是说，测试单元中心是否位于 S 中。

例程 scan_prim 执行隐式图元的外在型光栅化，并通过逐行（对应于图元包围盒单元的各行）扫描而实现。针对每个单元，将计算表面与射线（原点为 $(u,v,0)$，方向为 $(0,0,1)$）间的交点。如果存在交点，该单元随后将被例程 paint 处理。

```
230b <scan prim 230b>≡
  void scan_prim(Prim *p, void (*paint)())
  {
    Real u, v; Inode *l; Box3d bb = prim_bbox(p);

    for (v = bb.ll.y; v < bb.ur.y; v++)
      for (u = bb.ll.x; u < bb.ur.x; u++)
        if (l = prim_intersect(p, ray_make(v3_make(u,v,0), v3_make(0,0,1))))
          paint(u,v);
  }
Defines:
  scan_prim, never used.
```

需要注意的是，由于隐式图元定义于三维空间内，平行射线的交点对应于整合至光栅化中的正交投影。

考虑到可见性问题，三维场景的光栅化处理过程需要执行每个对象的同步光栅化。其中，将执行图像单元的扫描。针对每个单元，将计算对应射线与三维场景对象间的交点。如果多个对象与射线相交，则需要确定可见对象，该问题将在第 14 章中加以讨论。

13.4 基于细分的光栅化

本节将讨论直线段细分的光栅化方法。基于细分的光栅化算法由两项基本操作构成，即评估和分解。该处理过程始于初始配置。在每个递归级别，将进行评估以判断是否可枚举一个单元，或者是否有必要执行分解操作。对于后者，可能会执行细分操作，且算法递归地应用于各部分上。

13.4.1 内在型细分

直线段的内在型细分光栅化基于线段长度测试。如果测试结果小于单元直径，该线段将被例程 paint 所处理，否则将在中心点处细分线段，进而形成了递归操作。

例程 subdiv_line 实现了直线段的光栅化操作。

```
231a <subdiv line 231a>≡
  void subdiv_line(Vector3 p, Vector3 q, void (*paint)())
  {
    Box3d bb = bound(p, q);

    if ((bb.ur.x - bb.ll.x) <= 1 && (bb.ur.y - bb.ll.y) <= 1) {
      paint(bb.ll.x, bb.ll.y);
    } else {
      Vector3 m = v3_scale(0.5, v3_add(p,q));
      subdiv_line(p, m, paint);
      subdiv_line(m, q, paint);
    }
  }
Defines:
  subdiv_line, never used.
Uses bound 231b.
```

例程 bound 计算直线段的包围盒。

```
231b <bound 231b>≡
  static Box3d bound(Vector3 p, Vector3 q)
  {
    Box3d bb;
    bb.ll.x = MIN(p.x, q.x); bb.ll.y = MIN(p.y, q.y); bb.ll.z = MIN(p.z, q.z);
    bb.ur.x = MAX(p.x, q.x); bb.ur.y = MAX(p.y, q.y); bb.ur.z = MAX(p.z, q.z);
    return bb;
  }
Defines:
  bound, used in chunk 231a.
```

相应地，可对上述算法类型进行扩展，并执行多边形面片的光栅化操作，如 B 样条和 Bézier 曲面。其间将使用到现有的简单细分方法，其优点主要体现在效率和自适应性方面。

13.4.2 外在型细分

基于外在型细分的光栅化将划分包含于图像平面中的矩形，直至该矩形对应于某个网格单元，或者不再与光栅化对象相交。

例程 subdiv_boxline 实现了直线段的光栅化操作。

```
232a <subdiv boxline 232a>≡
 void subdiv_boxline(Box3d bb, Vector3 p, Vector3 q, void (*paint)())
 {

  if (disjoint(p, q, bb))
    return;

  if ((bb.ur.x - bb.ll.x) <= 1 && (bb ur.y - bb.ll.y) <= 1) {
    paint(bb.ll.x, bb.ll.y);
  } else {
    subdiv_boxline(b_split(bb, 1), p, q, paint);
    subdiv_boxline(b_split(bb, 2), p, q, paint);
    subdiv_boxline(b_split(bb, 3), p, q, paint);
    subdiv_boxline(b_split(bb, 4), p, q, paint);
  }
 }
Defines:
  subdiv_boxline, never used.
Uses b_split 233b and disjoint 232b.
```

例程 disjoint 计算线段 $\overline{pq}$ 是否与矩形 bb 相交。首先，例程测试线段是否完全包含于矩形边所定义的负半空间中，随后测试该矩形是否完全位于线段所确定的半空间内。

```
232b <disjoint 232b>≡
  static int disjoint(Vector3 p, Vector3 q, Box3d bb)
  {
    if (((p.x < bb.ll.x) && (q.x < bb.ll.x))|| ((p.y < bb.ll.y) &&
                                                (q.y < bb.ll.y)))
      return TRUE;
    if (((p.x > bb.ur.x) && (q.x > bb.ur.x))|| ((p.y > bb.ur.y) &&
                                                (q.y > bb.ur.y)))
      return TRUE;
```

```
    return same_side(p, q, bb);
  }
Defines:
  disjoint, used in chunk 232a.
Uses same_side 233a.
```

例程 same_side 计算盒体 bb 是否完全位于直线 $\overline{pq}$ 的同一侧。

```
233 <same side>233a≡
  static int same_side(Vector3 p,Vector3 q,Box3d bb)
  {
    Vector3 l=v3_cross(v3_make(p.x,p.y,1),v3_make(q.x,q.y,1));
    Real d0=v3_dot(l,v3_make(bb.ll.x,bb.ll.y,1));
    Real d1=v3_dot(l,v3_make(bb.ll.x,bb.ur.y,1));
    Real d2=v3_dot(l,v3_make(bb.ur.x,bb.ur.y,1));
    Real d3=v3_dot(l,v3_make(bb.ur.x,bb.ll.y,1));

    return((d0 < 0 && d1 < 0 && d2 < 0 && d3 < 0)
        || (d0 > 0 && d1 > 0 && d2 > 0 && d3 > 0));

  }
Defines:
  same_side,used in chunk 232b.
```

例程 b_split 执行矩形 b 的四元细分。

```
233b <bsplit 233b>≡
  static Box3d b_split(Box3d b, int quadrant)
  {
    switch (quadrant) {
    case 1: b.ll= v3_scale(0.5,v3_add(b.ll,b.ur)); break;
    case 2: b.ll.x = (b.ll.x + b.ur.x) * 0.5; b.ur.y
                  = (b.ll.y + b.ur.y) * 0.5; break;
    case 3: b.ur= v3_scale(0.5, v3_add(b.ll,b.ur)); break;
    case 4: b.ll.y = (b.ll.y + b.ur.y) * 0.5; b.ur.x
                  = (b.ll.x + b.ur.x) * 0.5; break;
    }
    return b;
  }
Defines:
  b_split, used in chunk 232a.
```

这一类型的算法可用于场景的光栅化和可见性计算，对应场景由包含异类描述的模型构成。此时，全部场景对象相对于 v 个矩形进行测试。该策略等同于经典的 Warnock

视见算法[Warnock 69b]。

13.5 补充材料

本章针对直线段、多边形和隐式图元介绍了光栅化算法。图 13.2 显示了基于内在型渐增式方法的多边形光栅化示例。

图 13.2 多边形光栅化

RASTER 库中的 API 涵盖了以下例程：

```
void scan_obj(Poly *p, void (*paint)(), void *rc);
Vector3 seg_bilerp(Poly *p, Real t, int lv, Real lt, int rv, Real rt);

void scan_space(Prim *p, void (*paint)());

void rsubdiv_obj(Vector3 p, Vector3 q, void (*paint)());

void rsubdiv_space(Box3d bb, Vector3 p, Vector3 q, void (*paint)());
```

13.6 本章练习

（1）编写一个程序，利用内在型细分光栅化一条直线段。
（2）编写一个程序，利用外在型细分光栅化一条直线段。
（3）编写一个程序，利用内在型渐增式光栅化方法光栅化一个多边形。

（4）编写一个程序，利用外在型渐增式光栅化方法光栅化一个多边形。

（5）尝试将练习（1）～（4）转换为交互式程序。其中，用户可指定光栅化的对象。

（6）利用直线光栅化程序绘制三维场景中的线框对象。

（7）编写一个程序，绘制隐式图元的轮廓。提示：可使用外在型渐增式光栅化方法的变化版本。

（8）针对 CSG 对象扩展练习（7）中的程序。

（9）针对光栅化操作，在对象中包含颜色属性。

（10）利用内在型细分并针对三角形开发一种光栅化方法。

第 14 章 可见表面计算

在视见处理过程中,三维场景表面投影至图像平面上。由于该投影不是双射的,因而有必要解决多个表面映射到同一像素时所产生的冲突问题。针对这一问题,本章将讨论可见性这一概念,并计算三维场景对象的可见表面。

14.1 基础知识

可见性问题实际上可归结为确定相机视域中的最近可见表面,这可视为一类排序问题。需要注意的是,这里仅关注部分排序,也就是说,沿着视见射线到达的第一个不透明表面。

可见表面计算与视见处理的其他操作关系紧密。可见性算法需要构建视见操作以实现其解决方案。当前上下文环境涉及以下各种关系:

- ❑ 视见转换应通过某种方式执行,对于管线中的各项计算,需要将对象置于适宜的坐标系中。
- ❑ 利用可见性算法,光栅化操作可通过多种方式加以整合。
- ❑ 一旦确定了可见表面,应执行光照计算,进而生成图像元素的颜色值。

14.1.1 场景属性和一致性

现有的各种场景属性构成了可见性算法的基础内容。据此,应对场景结构进行分析,并考查内部一致性以及复杂度关系。

当采用量化方式评估三维场景时,可采用多项标准,其中包括对象的数量、复杂性和同质性(相对大小)、场景中对象组的分布、对戏间的干涉,以及场景深度的复杂性。

一致性与场景对象的内在特征及其关系相关,进而决定了图像的变化程度。在可见性算法中,一致性类型基于以下内容:

- ❑ 对象。
- ❑ 面元和边。
- ❑ 图像。
- ❑ 场景深度。

- 时间变化。

需要注意的是，在具有较高复杂度的场景中，作为算法的效率因素，空间一致性的重要性有所降低。随着复杂度的增加，常见情形也将会发生逆转：单一像素中包含了对个对象，而不再是占据多个像素的单一对象。

14.1.2 表达和坐标系

可见性计算中需要考查的重点内容与对象几何体相关。可见性算法可具备通用特征或专有特性，进而接受场景对象的同质或异构几何描述。

可见性算法中最为常用的几何描述类型包括：
- 多边形网格。
- 双三次参数表面（样条、Bézier 曲线等）。
- 代数隐式表面（二次曲面、超二次曲面等）。
- 隐式构建实体几何模型（CSG）。
- 过程式模型（分形等）。

另一个与几何体相关的问题是视见系统采用的内部表达方式。相应地，通用性使得算法越发复杂，进而对性能产生负面影响。为了简化这一问题，某些视见系统会整合多个专有的过程，并单独执行某些操作。除此之外，还将生成一个可以在后续集成阶段组合的公共表示。这种策略的一个极端例子是，从一开始就将所有对象的几何形状转换为一种常见类型（例如，转换为多边形的近似结果）。另一种特殊情况是在最后一刻整合对象分组中的图像（如使用图像合成）。

可见性算法可分为世界空间操作和图像空间操作。第一个策略直接与对象的表达结果协同工作，经计算后生成投影至虚拟屏幕平面上的有序面元列表。其中可见性是所执行的第一项操作。这一算法类型通常采用参数描述。

工作于图像空间内的算法负责计算特定的分辨率水平（未必与图像保持一致）。其中，算法尝试针对每个像素处理相关问题，并沿视见射线方向分析相对深度。这里，可见性计算通常被推迟到最后一刻。该算法类型通常使用隐式描述。

14.1.3 分类

另外，可见性算法还可根据排序方法进行分类，从相机视角来看，此类方法用于确定可见表面。

可见性算法的排序结构与光栅化操作紧密相关，并根据场景对象确定图像区域。就可见性而言，占据图像非连续区域的对象可视为独立对象。同时，光栅化可以看作是一个有序的过程，在这个过程中，我们对每个对象占用的像素进行空间枚举。

实际上，光栅化处理将在方向 x 和 y 上（水平和垂直方向）生成有序结果，而可见性处理则沿 z 轴（深度）并在相机坐标系内生成有序结果。

可见性算法的计算结构采用了下列排序结构：(YXZ)、(XY)Z 和 Z(XY)。其中，括号表示组合排序操作。这 3 个排序结构对应于 3 种算法类型，同时结合光栅化、光栅化后的可见性以及光栅化之前的可见性处理可见性问题。

类型为(YXZ)和(XY)Z 的算法计算图像精度的可见性结果，并将排序问题归结为图像元素（像素）的邻接问题。类型为(YXZ)的算法在处理场景对象的同时求解可见性，相关示例包括 Z-缓冲区算法和扫描线算法。类型为 (XY)Z 的算法则针对每个像素整体求解可见性问题，如光线跟踪和递归细分算法。类型为 Z(XY)的算法将精确地计算可见性问题，同时采用全局方式处理 Z 排序（对象级别或对象面元级别），如 Z-排序、空间划分和递归剪裁算法。

14.2　Z-缓冲区

Z-缓冲区算法为每个像素存储图像中该点到最近表面的距离，即实际可见表面的距离。实际上，该算法对应于单元排序（桶排序）。

内部数据结构 **zbuf** 针对每个所需存储深度信息。

```
static Real *zbuf = NULL;
static int zb_h = 0, zb_w = 0;
```

宏 **ZBUF** 用于访问上述结构。

```
#define ZBUF(U,V) zbuf[U + V * zb_w]
```

例程 **zbuf_init** 通过分辨率($w \times h$)分配 Z-缓冲区。

```
240a <zbuff init 240a>≡
  void zbuf_init(int w, int h)
  {
    zbuf = (Real *) erealloc(zbuf, w * h * sizeof(Real));
    zb_w = w; zb_h = h;
    zbuf_clear(MAX_FLOAT);
  }
Defines:
  zbuf_init, never used.
Uses zb_h, zbuf, and zbuf_clear 240b.
```

例程 zbuf_clear 利用实际值初始化 Z-缓冲区。

```
240b <zbuff clear 240b>≡
  void zbuf_clear(Real val)
  {
    int x, y;
    for (y = 0; y < zb_h; y++)
      for (x = 0; x < zb_w; x++)
        ZBUF(x,y) = val;
  }
Defines:
  zbuf_clear, used in chunk 240a.
Uses zb_h and ZBUF.
```

例程 zbuf_store 确定点 $p = (x, y, z)$ 的可见性。如果其 z 坐标小于存储于 Z-缓冲区中像素(x, y)的距离，则该点处于可见状态。

```
240c <zbuff store 240c>≡
  int zbuf_store(Vector3 p)
  {
    int x = p.x, y = p.y;
    if ((x > 0 && y > 0 && x < zb_w && y < zb_h) && p.z < ZBUF(x,y)) {
      ZBUF(x,y) = p.z;
      return TRUE;
    } else {
      return FALSE;
    }
  }
Defines:
  zbuf_store, never used.
Uses zb_h and ZBUF.
```

例程 zbuf_peek 访问 Z-缓冲区，并返回存储于像素(x, y)中的距离值。

```
241a <zbuff peek 241a>≡
  Real zbuf_peek(Vector3 p)
  {
    int x = p.x, y = p.y;
    return (x > 0 && y > 0 && x < zb_w && y < zb_h)? ZBUF(x,y) : MAX_FLOAT;
  }
Defines:
  zbuf_peek, never used.
Uses zb_h and ZBUF.
```

14.3 光线跟踪

光线跟踪算法计算视见光线与三维场景中全部对象间的交点,并选取最近表面的交点。这一处理过程针对图像的每个像素予以执行,对应的光线驶离相机投影中心并穿越像素。

14.3.1 与三维场景对象的交点

光线跟踪方法计算光线与每个场景对象间的交点。其中,可见表面对应于沿光线方向、具有较小正参数 t 的交点。该过程涉及选择排序。

例程 ray_intersect 计算光线 r 与对象列表 olist 间的最近交点。

```
241b <ray intersect 241b>≡
  Inode *ray_intersect(Object *olist, Ray r)
  {
    Object *o; Poly *p;
    Inode *l = NULL, *i = NULL;
    for (o = olist; o !=  NULL; o = o->next ) {
      p = (o->type == V_POLYLIST)? o->u.pols : NULL;
      do {
        switch (o->type) {
        case V_CSG_NODE:
          l = csg_intersect(o->u.tcsg, r); break;
        case V_PRIM:
          l = prim_intersect(o->u.prim, r); break;
        case V_POLYLIST:
          if (p != NULL) {
            l = poly_intersect(p, poly3_plane(p), r);
            p = p->next;
          } break;
        }
        if ((l != NULL) && (i == NULL|| l->t < i->t)) {
          inode_free(i);
          i = l; i->m = o->mat;
          inode_free(i->next); i->next = NULL;
        }
      } while (p != NULL);
    }
```

```
    return i;
  }
Defines:
  ray_intersect, never used.
Uses csg_intersect 242.
```

14.3.2 与 CSG 模型间的交点

CSG 模型的光线跟踪方法计算与每个图元间的交点,并根据布尔操作对交点进行整合。该处理过程对应于合并排序。

例程 csg_intersect 计算光线与 CSG 实体间的交点。

```
242 <csg intersect 242>≡
  Inode *csg_intersect(CsgNode *n, Ray r)
  {
    if (n->type == CSG_COMP)
      return csg_ray_combine(n->u.c.op, csg_intersect(n->u.c.lft, r)
                                      , csg_intersect(n->u.c.rgt,r));
    else
      return prim_intersect(n->u.p, r);
  }
Defines:
  csg_intersect, used in chunk 241b.
Uses csg_ray_combine 243a.
```

根据布尔操作 op,例程 csg_ray_combine 整合列表 a 和 b 给出的 CSG 组合中两个元素的交点。

对应的交点是通过点列表得到的,其中,光线穿越以实体为界的表面。从这个列表开始,可沿光线确定区间,报告对应于实体的内、外点。此类区间可通过 CSG 操作加以整合,进而生成一个新的列表。

```
243a <csg ray combine 243a>≡
  Inode *csg_ray_combine(char op, Inode *a, Inode *b)
  {
    Inode in = {NULL, 0, {0,0,0}, 0, NULL}, *t, *head = *c = &in;
    int as = R_OUT, bs = R_OUT, cs = csg_op(op, as, bs);

    while (a|| b) {
      if ((a && b && a->t < b->t)|| (a && ! b))
        CSG_MERGE(as, a)
```

```
      else
        CSG_MERGE(bs, b)
    }
    c->next = (Inode *)0;
    return head->next;
  }
Defines:
  csg_ray_combine, used in chunk 242.
Uses CSG_MERGE 243b and csg_op 243c.
```

宏 CSG_MERGE 根据 CSG 操作合并两个区间。

```
243b <csg merge 243b>≡
  #define CSG_MERGE(S, A) { int ts = cs; \
    S = ! S; \
    if ((cs = csg_op(op, as, bs)) ! = ts) {\
      if (op == '-' && ! S) \
        A->n = v3_scale(-1., A->n); \
      c->next = A; c = A; A = A->next; \
    } else {\
      t = A; A = A->next; free(t); \
    }\
  }
Defines:
  CSG_MERGE, used in chunk 243a.
Uses csg_op 243c.
```

例程 csg_op 用于确定 CSG 操作结果。

```
243c <csg op 243c>≡
  int csg_op(char op, int l, int r)
  {
    switch (op) {
    case '+': return l| r;
    case '*': return l & r;
    case '-': return l & (! r);
    }
  }
Defines:
  csg_op, used in chunk 243.
```

光线跟踪方法的扩展包括随机采样、利用光线计算光照（光线跟踪）和优化操作。

14.4　Painter 算法

Painter 算法也称作 Z-排序，该算法分为两个阶段。在第一个阶段中，场景组件相对于虚拟相机进行排序；在第二个阶段中，对象从最远到最近按顺序光栅化。

Painter 算法包含两种可能的实现方式，即近似 Z-排序方法和完全 Z-排序方法。

14.4.1　近似 Z-排序

在近似 Z-排序方法中，多边形根据三角形与观察者间的距离值加以存储。该值可从形心位置，甚至是多边形的某个顶点处进行估算。另外，从可见表面计算角度来看，仅对每个多边形使用距离值无法保证多边形的正确排序。然而，该方法易于实现，且适用于大多数场合，同时形成了完全 Z-排序方法的初始阶段。

数据结构 Zdatum 存储了对象的多边形表达值。

```
244a <zdatum 244a>≡
  typedef struct Zdatum {
    Real zval;
    Poly *l;
    IObject *o;
  } Zdatum;
Defines:
  Zdatum, used in chunk 245.
```

例程 z_sort 使用排序插入方法，并根据 z 中的距离构建场景多边形列表。

```
244b <zsort 244b>≡
  List *z_sort(List *p)
  {
    List *q = new_list();
    while (! is_empty(p)) {
      Item *i = z_largest(p);
      remove_item(p, i);
      append_item(q, i);
    }
    return q;
  }
Defines:
  z_sort, never used.
Uses z_largest 245.
```

例程 z_largest 返回较大 z 坐标的列表元素 p。

```
245  <zlargest 245>≡
  Item *z_largest(List *p)
  {
    Item *i, *s;

    for (s = i = p->head; i != NULL; i=i->next)
      if (((Zdatum *)(i->d))->zval > ((Zdatum *)(s->d))->zval)
        s = i;
    return s;
  }
Defines:
  z_largest, used in chunk 244b.
Uses Zdatum 244a.
```

14.4.2 完全 Z-排序

完全 Z-排序方法始于近似 Z-排序生成的有序多边形列表。随后，该算法遍历这一列表，并判断多边形顺序相对于可见表面计算的视点是否正确。如果两个多边形 P 和 Q 未包含正确的顺序，则需要调整它们在列表中的位置。

当判断绘制顺序是否正确时，该方法要求多边形 Q（在多边形 P 之后被绘制）不能产生遮挡。该过程可通过一系列的测试完成，同时也增加了计算复杂度，如测试 P 和 Q 的包围盒、通过其他支撑平面划分多边形、屏幕上 P 和 Q 投影的交点。

某些时候，多边形序列可能会形成一个与绘制标准相关的循环。对于这种情况，通常无法确定一个正确的顺序，因此有必要通过细分一个多边形来打破这个循环。

算法 14.1 中显示了完全 Z-排序的伪代码。

```
sort l by the centroid in z (approximated Z-sort);
while l ≠ ∅ do
  select P and Q;
  if P does not occlude Q then
    continue;
  else if Q flagged then
    resolve cycle;
  else if Q does not occlude P then
    swap P with Q;
    flag Q;
  end if
```

```
end while
paint l in order;
```

算法 14.1 full_zsort(*l*)

14.5 其他可见性方法

本节将展示一些当前系统中未采用的其他可见性算法。

14.5.1 空间细分

细分算法根据虚拟相机的参数对场景对象进行单独分类，同时创建一个数据结构。一旦相机位置确定后，即可通过某种方式遍历该数据结构，进而表示正确的对象可见顺序。

该结构一般采用二叉空间划分，或 BSP 树。相关方法包含以下两个步骤：

（1）预处理。构建 BSP 树。

（2）可见性。根据相机遍历该结构。

算法 14.2 描述了构造例程 make_bsp；算法 14.3 所示的 bsp_traverse 例程通过 BSP 树计算可见性问题。

```
if plist == NULL then
  return NULL;
end if
root = select(plist);
for all p ∈ plist do
  if p on '+' side of the root then
    add(p, frontlist);
  else if p on '-' side of the root then
    add(p, backlist);
  else
    split_poly(p, root, fp, bp);
    add(fp, frontlist);
    add(bp, backlist);
  end if
end for
return combine(make_bsp(frontlist),make_bsp(backlist));
```

算法 14.2 BSP make_bsp(plist)

```
if t == NULL then
  return;
end if
if c in front oft.root then
  bsp_traverse(c, t->back);
  render(t->root);
  bsp_traverse(c, t->front);
else
  bsp_traverse(c, (t->front);
  if backfaces then
    render(c, t->root);
  end if
  bsp_traverse(c, t->back);
end if
```

算法 14.3 bsp_traverse(c, t)

14.5.2 递归细分

递归细分算法将图像递归分解为 4 个象限，直到该区域的场景配置具有一个简单的可见性求解结果，或者该区域的大小为像素。递归细分算法也称作 Warnock 算法[Warnock 69b]。算法 14.4 描述了例程 recursive_subdivision。

```
for p in plist do
  if P in r then
    classify P;
  else
    remove P from plist;
  end if
end for
if configuration == SIMPLE then
  render r;
else
  divide r into 4;
  recursive_subdivision(plist, quadrant 1);
  recursive_subdivision(plist, quadrant 2);
  recursive_subdivision(plist, quadrant 3);
  recursive_subdivision(plist, quadrant 4);
end if
```

算法 14.4 recursive_subdivision(plist, r)

```
if list == empty then
  return;
end if
sort approximately in Z;
select front polygon P;
divide list into inside e outside clipping in relation to P;
while inside != empty do
  select Q;
  if P in front of Q then
    remove Q;
  else
    swap P and Q;
  end if
end while
render P;
recursive_clipping(outside);
```

算法 14.5 recursive_clipping(lista)

14.6 补充材料

[Sutherland et al. 74]首次系统分析了计算可见表面的算法，提出了一种基于排序准则的可见性问题表征方法。

针对异构几何场景的可见性问题，存在多种解决方案。其中，[Crow 82]针对对象分组执行可见性计算，并通过图像合成在后处理过程中对其进行整合。[Cook et al. 87]通过预处理将所有场景表面转换为微多边形，进而执行高效的可见性计算。

[Newell et al. 72a]和[Newell et al. 72b]描述了 Z-排序算法。

[Shumacker et al. 69]首次在飞行模拟应用程序中提出了空间划分算法；随后，[Fuchs et al. 83]将其应用扩展至更加通用的应用程序中。

Weiler 和 Atherton 在计算可见表面时提出了递归剪裁算法（[Weiler and Atherton 77]，[Atherton et al. 78]）。[Heckbert and Hanrahan 84]对此进行了改进，并应用于光束跟踪算法中。

[Carpenter 84]发布的 A-缓冲区算法对 Z-缓冲区算法进行了扩展，并在子像素级别实现了可见表面的更为精准的计算。

[Warnock 69a]提出了递归图像细分算法；扫描线可见性算法则是由[Watkins 70]和[Bouknight and Kelly 70]独立开发的。

图 14.1 显示了基于 Painter 算法的可见表面计算结果。图 14.2 显示了包含同一场景中

Z 值的图像结果。

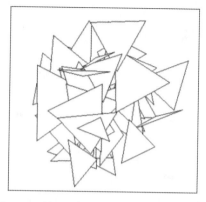

图 14.1 基于 Painter 算法的可见表面计算

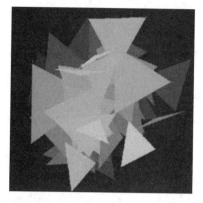

图 14.2 深度值（Z-缓冲区）

可见表面计算库中的 API 包含以下例程：

```
void zbuf_init(int w, int h);
void zbuf_clear(Real val);
int zbuf_store(Vector3 p);

Inode *ray_cast(Object *olist, Ray r);

Inode *csg_intersect(CsgNode *n, Ray r);
Inode *csg_ray_combine(char op, Inode *a, Inode *b);
int csg_op(char op, int l, int r);

List *z_sort(List *p);
Item *z_largest(List *p);

Item *new_item(void *d);
List *new_list();
int is_empty(List *q);
void append_item(List *q, Item *i);
void remove_item(List *q, Item* i);
void free_list(List *q);
```

14.7 本章练习

（1）编写一个程序并生成一组三角形。使用下列参数作为输入：三角形数量、尺寸、方向、宽高比、相对于主方向(x, y, z)的分布、颜色值。输出结果应为 SDL 语言中的三角形列表。

（2）编写一个程序并生成一组球体。输入参数与练习（1）相同。输出结果应为 SDL 语言中的一组图元球体。

（3）编写一个程序扫描 SDL 图元，并生成一个与其近似的多边形网格。输出结果应为 SDL 语言中的三角形列表。

（4）编写一个程序，利用 Z-缓冲区计算可见性结果。输入内容为练习（1）生成的三角形列表。输出应为：① 包含可见表面的图像；② 包含 Z 值的图像。

（5）编写一个程序，利用光线跟踪计算可见性结果。输入内容为练习（2）中的图元列表。输出应为：① 包含可见表面的图像；② 包含 Z 值的图像。

（6）编写一个程序，利用 Painter 算法计算可见性问题。输入内容为练习（1）生成的三角形列表。输出应为：① 包含可见表面的图像；② 包含 Z 值的图像。

（7）针对同一球体集合，比较上述练习中的结果。情况 1：球体间不相交；情况 2：球体间相交。

（8）针对一组三角形，重复练习（7）中的比较操作。

第 15 章 局部光照模型

当颜色信息在图像中表示的对象上设置完毕后,视见过程即处理完毕。在 3D 场景中,这种对应关系可以通过光照计算完成。由于模拟了物理环境的视觉感受,因而也是一种较为自然的选择方案。本章和后续章节将讨论三维场景可视化过程中的光照和着色处理过程。

15.1 基础知识

光照和着色可通过 4 种环境范例予以理解。在物理世界中,光照与光线和物质间的交互有关。在数学环境下,一些方法可模拟光照模型。在实现过程中,某些计算模式可定义着色函数,如图 5.1 所示。

15.1.1 光照

三维场景由发光或传输能量的对象构成。在视见处理过程中,我们使用了虚拟相机,并将其视为一种光线敏感设备。

光线包含了两种特征,即粒子束或波。光线的粒子模型假定沿着一条射线的能量流是由运动中的粒子(称为光子)量化的,该模型是在几何光学中研究的。另一方面,光的波动模型通过两个场的结合来描述光能,即电场和磁场,并通过电磁学中的麦克斯韦方程加以描述。

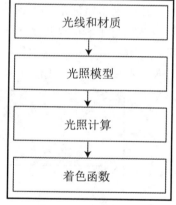

图 15.1 光照的抽象级别

另外,一些光照现象可通过粒子模型加以解释,而另外一些则通过波动模型定义。

在计算机图形学中,光线的粒子模型较为常见,其原因在于,可通过简单方式模拟三维场景光照中的大多数现象。本书主要讨论粒子模型,且仅关注光照表面所涉及的问题。这意味着,介质并不参与光照计算,光线将在真空中传播。上述选择结果可极大地简化当前问题。

15.1.2 光线传播

光照主要研究光在环境空间中的传播。当理解各项传播中所涉及的各种概念时，应该区分辐射度和光照度。这里，辐射度是指辐射能的发射，并用于光照计算中；而光照度则指辐射能的感知，主要用于着色函数中。

通过电磁波形式传播的能量称作辐射能。辐射束一般由可见光谱中的各种波构成。将每种辐射波长与对应能量关联的函数称作光谱分布。

我们将单位时间内由表面发射、传输或接收的辐射能称作辐射流。辐射流是一个标量值。当考查辐射能的传播时，应区分特定方向上表面的入射流（称作辐照度）和出射流（称作辐射强度或辐射度）。

光流是通过计算观测者所感知的辐射能而得到的量值；标准感受器的灵敏度则通过发光效率函数予以测量。

辐射能沿直线传播，当到达两种介质的界面时，部分能量将被反射，即返回至原传播介质中；而另一部分将被传输，也就是说，穿越表面并进入另一种媒介。

几何光学定理定义了到达表面的光线路径。其中，反射定律表明，入射光线 I 和反射光线 R 相对于表面法线 N 形成了相等的角度，即 $\theta_i = \theta_r$。而传输过程则遵循斯涅耳定律，即 $n_I \sin\theta_i = n_T \sin\theta_t$。其中，$n_I$ 和 n_T 表示为由表面分隔的介质的折射系数；θ_I 和 θ_T 分别表示入射光线 I 和传输光线 T 相对于表面法线 N 形成的角度，如图 15.2 所示。

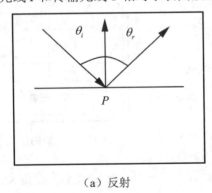
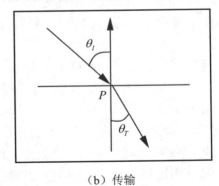

(a) 反射　　　　　　　　　　　　(b) 传输

图 15.2　反射和传输

15.1.3 表面和材质

我们主要关注两种介质边界处的光线和材质的交互行为。这里，该交互行为取决于表面几何体以及对象的材质,这将最终确定辐射能量及其在每次交互过程中的传播路径。

材质可分为绝缘体、金属和复合材料。其中,绝缘材质通常是半透明的,并用作导电绝缘体,如玻璃。金属材质一般为不透明物体且用作电导体,如铜、金、铝。复合材质则是由悬浮于透明基质中的不透明颜料形成的,如塑料和涂料。

从光学角度上来看,表面的几何形状可以是光滑的或者是粗糙的。光滑表面可通过其切平面实现局部建模。在光滑表面中,光线沿反射或传输方向传播。相比之下,粗糙表面则通过朝向各个方向上的微面元模型予以近似。在粗糙表面中,光线沿多个方向传播,如图 15.3 所示。

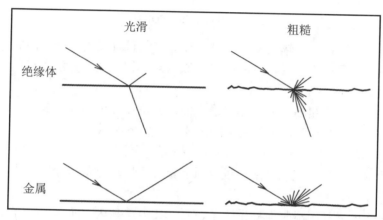

图 15.3　光滑和粗糙表面上的光线传输和反射

能量守恒则是光线传播行为的一个基本原理,表明到达两个介质间表面的能量将被反射、传输或吸收。换而言之,反射、传输和吸收能量之和等于入射能量。

在空气和绝缘体材质间的边界处,大多数光线将处于传输状态;在空气和金属材质间的边界处,大多数光线将被反射。除此之外,某些光能在相互作用中被吸收了(如转换为热能)。总体来说,反射和传输光线的比例取决于入射光线和表面法线间的角度。

15.1.4　局部光照模型

表面的局部光照模型描述了两种介质间界面处的交互结果。该模型应可预测辐射能的传播路径,以及源自入射光线光谱分布,进而模拟环境空间内表面点之间的光能传输。

双向反射分布函数(BRDF)和双向透射率分布函数(BTDF)可用于创建局部光照模型。此类函数取决于入射光线的方向和辐射能量,并在表面光线入射点处确定了沿半球各方向上的最终辐射能量。图 15.4 显示了既定入射方向上的 BRDF 和 BTDF 函数。

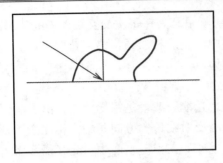
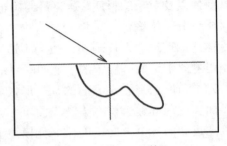

(a) BRDF 函数　　　　　　　　(b) BTDF 函数

图 15.4　BRDF 和 BTDF 函数

反射和传输函数可通过辐射度量值进行计算，其过程较为复杂。因此，这里将通过简单函数对理想表面行为建模，进而实现近似结果。相应地，光滑和粗糙表面可视为两种较为极端的示例。

理想状态下的光学粗糙表面称作郎伯表面，并遵循郎伯定律。其中，任意方向上的入射能量正比于入射光线和法线间夹角的余弦值，如下所示：

$$E_d = E_i \cos\theta$$

这里，反射呈漫反射状态，或者说反射能量在光照半球的各个方向上是恒定的，如图 15.5（a）所示。

理想状态下的光学平滑表面称作镜面表面。其中，到达表面点 p 处的光线沿方向 R 反射，且入射和反射角相等，如图 15.2 所示。

由于能量在方向 R 上反射，因而它表示为点 p 处到达光照半球的全部能量。在大多数材质中，镜面反射集中在反射方向上，并在远离方向上逐渐减弱。镜面反射的经验光照模型则是 Phong 模型[Phong 75]。根据该模型，在方向 V 上反射的辐射变量正比于夹角 β（V 和反射方向 R 间的夹角）余弦值的 p 次方，如图 15.5（b）所示。

$$E_s = E_i (\cos\beta)^p$$

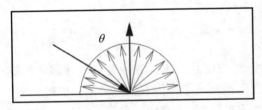
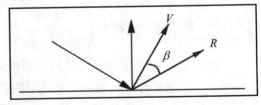

(a) 漫反射　　　　　　　　(b) 镜面反射

图 15.5　漫反射和镜面反射

漫反射和镜面光照的局部模型经整合后可近似模拟表面的双向反射函数，如下所示：

$$E_o = k_d E_d + k_s E_s$$

其中，k_d 和 k_s 表示为与材质相关的常量。

其他高级光照模型还包括 Blinn[Blinn and Newell 76]、Cook-Torrance[Cook and Torrance 81]和 Torrance-Sparrow [Torrance and Sparrow 76]模型。

15.2 光　　源

本节将讨论光源的实现和光能传输机制。此类机制也是实现局部光照方程的基本内容。

15.2.1 光线传输

当实现光线传输时，可定义一个新的几何元素，即光锥，如图 15.6 所示。结构 Cone 描述了一个光锥，其中包含了光源 o、方向 d 以及锥体扩展角 α 的余弦值。

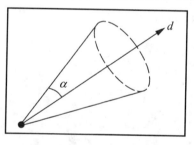

图 15.6　光锥

```
256a <cone 256a>≡
  typedef struct Cone {
    Vector3 o, d;
    Real cosa;
  } Cone;
Defines:
  Cone, used in chunks 256-61.
```

例程 cone_make 实现了 Cone 的构造函数。

```
256b <cone make 256b>≡
  Cone cone_make(Vector3 o, Vector3 d, Real angle)
  {
    Cone c;
```

```
    c.o = o; c.d = d; c.cosa = cos(angle);
    return c;
  }
Defines:
  cone_make, used in chunks 260b and 261.
Uses Cone 256a.
```

例程 dir_coupling 和 point_coupling 实现了锥体的可见性计算，并用于光线传输的几何计算中，如图 15.7 所示。

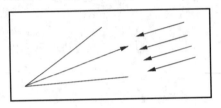
（a）锥体和方向间的可见性

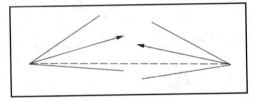
（b）两个锥体间的可见性

图 15.7　可见性关系

例程 dir_coupling 确定锥体 C 的传播方向是否可见。

```
256c <dir coupling 256c>≡
  int dir_coupling(Cone a, Vector3 v)
  {
    if (v3_dot(a.d, v3_scale(-1, v)) > a.cosa)
      return TRUE;
    else
      return FALSE;
  }
Defines:
  dir_coupling, used in chunks 257-59.
Uses Cone 256a.
```

例程 point_coupling 用于确定两个锥体间是否彼此可见。

```
257 <point coupling 257>≡
  int point_coupling(Cone a, Cone b)
  {
    Vector3 d = v3_unit(v3_sub(a.o, b.o));
    return dir_coupling(a, d) && dir_coupling(b, v3_scale(-1, d));
  }
Defines:
  point_coupling, never used.
Uses Cone 256a and dir_coupling 256c.
```

15.2.2 光源的表达

相应地，存在多种光源类型。其中，最为常见的是有向光源、点光源和聚光灯。有向光源通常距离遥远，如太阳，其能量在单一方向上传播，且不会随着距离而衰减。点光源则是一类局部光照，如蜡烛或白炽灯，其能量在各个方向上传播，并随距离而衰减。聚光灯则是一类局部集中光源，如台灯或剧场反光灯。

这里将采用结构 Light 表示一个光源，该结构定义为单链表，并存储各种光源类型的参数。接下来将描述各种光源的公共元素，其中，type 定义为光源类型；color 指定了光线的光谱分布；intensity 表示为光强；transport 指向一个函数，该函数实现了光线传输；outdir 表示为表面交互计算得到的光线发射方向；outcol 表示交互过程中发射的能量分布。

```
258a <light struct 258a>≡
  typedef struct Light {
    struct Light *next;
    int          type;
    Color        color;
    Real         ambient;
    Real         intensity;
    Vector3      loc;
    Vector3      dir;
    Real         cutoff;
    Real         distr;
    Real         att0, att1, att2;
    Vector3      outdir;
    Color        outcol;
    int          (*transport)();
    void         *tinfo;
  } Light;
Defines:
  Light, used in chunks 258-61, 263a, and 310c.
Uses ambient 260a.
```

例程 ambientlight 实现了环境光传输。

```
258b <ambient light 258b>≡
  int ambientlight(Light *l, Cone recv, RContext *rc)
  {
    return FALSE;
  }
Defines:
```

ambientlight, never used.
Uses Cone 256a, Light 258a, and RContext 259b.

例程 distantlight 实现了有向光源的传输。

```
258c <distant light 258c>≡
  int distantlight(Light *l, Cone recv, RContext *rc)
  {
    if (dir_coupling(recv, l->dir) == FALSE)
      return FALSE;
    l->outdir = v3_scale(-1, l->dir);
    l->outcol = c_scale(l->intensity, l->color);
    return TRUE;
  }
Defines:
  distantlight, used in chunk 263a.
Uses Cone 256a, dir_coupling 256c, Light 258a, and RContext 259b.
```

例程 pointlight 实现了点光源的传输。

```
259a <point light 259a>≡
  int pointlight(Light *l, Cone recv, RContext *rc)
  {
    Real d, dist, atten;
    Vector3 v = v3_sub(rc->p, l->loc);
    dist = v3_norm(v);
    l->dir = v3_scale(1/dist, v);
    if (dir_coupling(recv, l->dir) == FALSE)
      return FALSE;
    atten = ((d = l->att0 + l->att1*dist + l->att2*SQR(dist))) > 0)? 1/d : 1;
    l->outdir = v3_scale(-1, l->dir);
    l->outcol = c_scale(l->intensity * atten, l->color);
    return TRUE;
  }
Defines:
  pointlight, never used.
Uses Cone 256a, dir_coupling 256c, Light 258a, and RContext 259b.
```

15.3 局 部 光 照

本节将讨论局部光照模型的实现。

15.3.1 光照上下文

局部光照计算相对于场景表面中的一点进行。光照函数定义于该点的光照半球中，表示为点 p 处表面切平面上的光线集合，如图 15.8 所示。

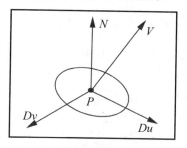

图 15.8 局部光照几何体

结构 RContext 存储了局部光照所用的数据。

```
259b <render context 259b>≡
  typedef struct RContext {
    Vector3 v, p, n;
    Vector3 du, dv;
    Vector3 t;
    Material *m;
    Light *l;
    View *c;
    Image *img;
  } RContext;
Defines:
  RContext, used in chunks 258-62, 304-6, and 310.
Uses Light 258a and Material 262a.
```

15.3.2 光照函数

这里将根据环境光、漫反射光和镜面光使用简单的局部光照模型。需要注意的是，全部计算均在基于圆锥体 receiver（扩展角 $\alpha = \pi/2$）的例程中所描述的光照半球上执行。例程 ambient 实现了环境光，并计算环境中的恒定光照。

```
260a <ambient 260a>≡
  Color ambient(RContext *rc)
  {
```

```
    Light *l; Color c = C_BLACK;
    for (l = rc->l; l ! = NULL; l = l->next)
        c = c_add(c, c_scale(l->ambient, l->color));
    return c;
  }
Defines:
  ambient, used in chunks 258a and 262b.
Uses Light 258a and RContext 259b.
```

例程 diffuse 遵循朗伯定律,并实现了漫反射光。每种光源的贡献结果表示为 $c_i<N, L_i>$,其中,N 表示为表面法线,L_i 和 c_i 分别表示为光源 i 的入射方向和能量分布(在例程 diffuse 中,L_i 和 c_i 通过 l-> outdir 和 l-> outcol 指定)。

```
260b <diffuse 260b>≡
  Color diffuse(RContext *rc)
  {
    Light *l; Color c = C_BLACK;
    Cone receiver = cone_make(rc->p, rc->n, PI/2);

    for (l = rc->l; l ! = NULL; l = l->next)
      if ((*l->transport)(l, receiver, rc))
        c = c_add(c, c_scale(v3_dot(l->outdir, rc->n), l->outcol));
    return c;
  }
Defines:
  diffuse, used in chunk 262b.
Uses Cone 256a, cone_make 256b, Light 258a, and RContext 259b.
```

例程 specular 采用了 Blinn 模型实现了镜面光照,该模型类似于 Phong 模型,同时避免了计算反射向量 R,因而更加高效。计算过程是以向量 $H = (L + V)/2$ 的函数这一形式而进行的,即向量 L(光源方向)和 V(观察者方向)的等分线。光照项通过 $c_i<H_i, N>^{s_e}$ 给出。其中,N 表示为表面的法线;c_i、H_i 和 s_e 则分别对应于能量分布、光源 i 的等分向量以及镜面反射指数。

```
261 <specular 261>≡
  Color specular(RContext *rc)
  {
    Light *l; Vector3 h; Color c = C_BLACK;
    Cone receiver = cone_make(rc->p, rc->n, PI/2);

    for (l = rc->l; l ! = NULL; l = l->next) {
      if ((*l->transport)(l, receiver, rc)) {
```

```
            h = v3_unit(v3_scale(0.5, v3_add(l->outdir, rc->v)));
            c = c_add(c, c_scale(pow(MAX(0, v3_dot(h, rc->n)),rc->m->se),
                  l->outcol));
       }
    }
    return c;
}
Defines:
  specular, used in chunk 262b.
Uses Cone 256a, cone_make 256b, Light 258a, and RContext 259b.
```

15.4 材质

本节将讨论光照计算中材质的实现过程。

15.4.1 描述材质

结构 Material 用于描述材质，并存储了局部光照计算中所采用的材质参数。其中，颜色 C 和 S 指定了材质的漫反射和镜面反射的光谱分布；常量 k_a、k_d、k_s 分别表示为环境反射、漫反射和镜面反射系数。常量 s_e 表示为镜面反射系数，常量 k_t 表示透射系数，i_r 则表示为材质的折射系数。另外，指针 luminance 指向实现了材质行为的例程。该例程一般基于局部光照模型。

```
262a <material struct 262a>≡
  typedef struct Material {
     Color c, s;
     Real ka, kd, ks, se, kt, ir;
     Color (*luminance)();
     void *tinfo;
  } Material;
Defines:
  Material, used in chunks 259b, 263b, 304-6, and 310c.
```

15.4.2 材质类型

如前所述，材质可分为绝缘体、金属和复合材料。本节将讨论单一材质类型的实现，即塑料。这是一类复合材质，并采用了基于漫反射和镜面光照的光照模型。

例程 plastic 实现了塑料材质，该类型的光照方程如下所示：
$$C(k_a \operatorname{amb}(p) + k_d \operatorname{diff}(p)) + S(k_s \operatorname{spec}(p))$$

其中，amb、diff 和 spec 分别表示环境光、漫反射光和镜面光照项，并在点 p 处的光照上下文 rc 中予以计算。

```
262b <plastic 262b>≡
 Color plastic(RContext *rc)
 {
    return c_add(c_mult(rc->m->c, c_add(c_scale(rc->m->ka, ambient(rc)),
                                        c_scale(rc->m->kd, diffuse(rc)))),
                 c_mult(rc->m->s, c_scale(rc->m->ks, specular(rc))));
 }
Defines:
  plastic, never used.
Uses ambient 260a, diffuse 260b, RContext 259b, and specular 261.
```

15.5 语言规范

例程 distlight_parse 实现了 SDL 语言中的有向光源。

```
263a <parse distlight 263a>≡
 Val distlight_parse(int pass, Pval *pl)
 {
   Val v;
   if (pass == T_EXEC) {
     Light *l = NEWSTRUCT(Light);
     l->type = LIGHT_DISTANT;
     l->color = C_WHITE;
     l->intensity = pvl_get_num(pl, "intensity", 1);
     l->dir = v3_unit(pvl_get_v3(pl, "direction", v3_make(1,1,1)));
     l->transport = distantlight;
     v.type = LIGHT;
     v.u.v = l;
   }
   return v;
 }
Defines:
  distlight_parse, never used.
Uses distantlight 258c and Light 258a.
```

例程 metal_parse 实现了 SDL 语言中的金属材质。

```
263b <parse metal 263b>≡
  Val metal_parse(int pass, Pval *pl)
  {
    Val v;
    if (pass == T_EXEC) {
      Material *m = NEWSTRUCT(Material);
      m->c = pvl_get_v3(pl, "d_col", C_WHITE);
      m->ka = pvl_get_num(pl, "ka", .1);
      m->ks = pvl_get_num(pl, "ks", .9);
      m->se = pvl_get_num(pl, "se", 10);
      m->luminance = metal;
      v.type = MATERIAL;
      v.u.v = m;
    }
    return v;
  }
Defines:
 metal_parse, never used.
Uses Material 262a.
```

15.6 补充材料

图 15.9 显示了执行光照计算的交互式程序。

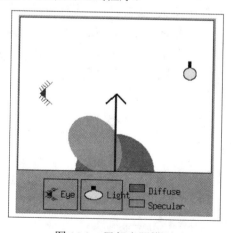

图 15.9 局部光照模型

SHADE 库中的 API 包含以下例程。

```
Cone cone_make(Vector3 o, Vector3 d, Real angle);
int point_coupling(Cone a, Cone b);
int dir_coupling(Cone a, Vector3 v);

Vector3 faceforward(Vector3 a, Vector3 b);
RContext *rc_set(RContext *rc, Vector3 p, Vector3 n, Vector3 v);

Color ambient(RContext *rc, Light *l, Material *m, Vector3 d);
Color diffuse(RContext *rc, Light *l, Material *m, Vector3 d);
Color specular(RContext *rc, Light *l, Material *m, Vector3 d);

int ambientlight(Light *l, Cone recv, RContext *rc);
int distantlight(Light *l, Cone recv, RContext *rc);
int pointlight(Light *l, Cone recv, RContext *rc);
int spotlight(Light *l, Cone recv, RContext *rc);

Color constant(RContext *rc, Light *l, Material *m);
Color matte(RContext *rc, Light *l, Material *m);
Color metal(RContext *rc, Light *l, Material *m);
Color plastic(RContext *rc, Light *l, Material *m);
```

15.7 本章练习

（1）编写一个程序，针对可变的光源位置，计算表面某点处的漫反射光照。

（2）编写一个程序，针对可变的光源和观察者位置，计算表面某点处的镜面光照。

（3）将练习（1）和（2）整合至交互式图形程序中。其中，用户可调整光源和观察者的位置。程序应可通过可视化方式显示光照计算结果。

第 16 章 全 局 光 照

三维场景光照表示为光源与环境表面间的全局交互结果。本章将讨论描述这一现象的光照方程以及计算方法,并采用数值方式求解光照方程。

16.1 光照模型

光照方程阐述了辐射能量在环境空间内的传播。如前所述,根据粒子光照模型,每个光子均加载一定量的辐射能量。在某一特定时刻,光子可通过其位置和运动方向定义。通过这一方式,光子的状态由 (s,ω) 确定。其中,$s \in \mathbb{R}^3$,$\omega \in S^2$,空间 $\mathbb{R}^3 \times S^2$ 称作相空间。

光照表示为光子运动结果。因此,需要计算单位时间内的辐射能量流,或辐射功率。这里辐射流 $\Phi(s,\omega,t)$ 定义为相空间内的函数,如下所示。

$$\Phi(s, \omega, t)\mathrm{d}A\mathrm{d}\omega$$

该函数表示为穿越点 s 邻接区域内微分面积 $\mathrm{d}A$ 的光子数量,且时刻 t、方向 w 邻接区域的微分立体角为 $\mathrm{d}w$,如图 16.1 所示。

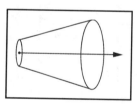

图 16.1 辐射能量流的几何表达

光照方程是传输理论中的一部分内容,主要描述时空内抽象粒子的分布状态。该理论提出两个假设条件,以适应于表面光照的研究,从而极大地简化了光照问题。

在第一个假设条件中,系统处于守恒状态。换而言之,环境中的光照传播条件不会在模拟时间内发生变化。该条件意味着,能量流在场景中的每一点均保持恒定,因而有:

$$\frac{\partial \Phi}{\partial t} = 0$$

第二个假设条件表示传输过程在真空中进行,因而所有行为均出现于场景对象表面上。

16.1.1 传输过程

真空中两点之间的辐射能量传输可通过下列方程定义：

$$\Phi(r, \omega) = \Phi(s, \omega) \quad (16.1)$$

其中，r、s 表示为沿 w 方向上彼此可见的两点。需要注意的是，该方程仅对满足可见条件的、环境空间内的点对有效。这里仅关注场景的表面点 r、$s \in M = \cup M_i$。

当给定点 r 后，可利用表面可见函数 $v: \mathbb{R}^3 \times S^2 \to \mathbb{R}$ 计算点 $s \in M$，并返回场景表面上最近可见点的距离，如下所示：

$$v(r, \omega) = \inf\{\alpha > 0 : (r - \alpha\omega) \in M\}$$

随后，点 s 可通过 $s = r - v(r, \omega)\,\omega$ 得到。

表面上点 p 处的光照方程取决于各方向到达的辐射能量流。对此，可使用点 p 处的光照半球 Θ 这一概念，即中心位于点 p、半径为 1 的虚球体。图 16.2 显示了两个表面可见点间的辐射能量传输几何描述。

源自点 p 的、方向 w 上的光照能量定义了半球 Θ_o 的实体角；相反，到达点 p 处的光照能量则定义了光照半球 Θ_i 上的实体角。通过这种方式，光照半球将控制表面点 p 和环境间的能量交换，如图 16.3 所示。

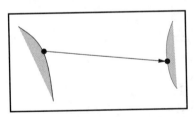

图 16.2 两个表面可见点间的辐射能量传输

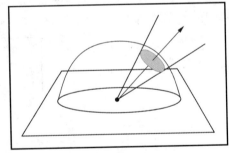

图 16.3 光照半球

16.1.2 边界条件

式（16.1）描述了真空中辐射能量的传输条件。为了完整地模拟光照问题，还需要指定边界条件。换而言之，当光线到达场景表面时，情况又当如何？对此，存在以下两种边界条件：

- 显式。离开表面点 s 的能量流（方向为 w）与入射能量流无关。对应的光照函数如下所示。

第 16 章 全局光照

$$\Phi(s,\omega) = \varepsilon(s,\omega) \tag{16.2}$$

其中，ε 指定了表面的发射函数。

- 隐式。离开表面点 s 的能量流（方向为 ω）取决于光照半球的入射流，如下所示。

$$\Phi(s,\omega) = \int_{\Theta_i} k(s,\omega' \to \omega)\Phi(s,\omega')\mathrm{d}\omega' \tag{16.3}$$

其中，$\omega \in \Theta_o$，且 $\omega \in \Theta_i$，函数 k 表示为表面的双向反射函数。需要注意的是，考虑到能量守恒定律，驶离 Θ_o 的辐射能量须小于 Θ_i 中的入射辐射能量。

16.1.3 辐射度方程

下面考查定义了传输方程中边界条件的光照方程。

在每一点 $r \in M$ 处，需要获得源自各点 $s \in M$（在 r 中的光照半球 Θ_i 内可见）的辐射能量贡献结果。

对此，可将两个点 r 和 s 之间的传输方程划分为：

$$\Phi(r,\omega) = \Phi(s \to r,\omega) \tag{16.4}$$

并替换等式右侧的两个边界条件，即式（16.2）和式（16.3），如下所示。

$$\Phi(r,\omega) = \varepsilon(s,\omega) + \int_{\Theta_i} k(s,\omega' \to \omega)\Phi(s,\omega')\mathrm{d}\omega' \tag{16.5}$$

传输方程根据光子数量（辐照度）描述了辐射能量流。然而，此处需要得到辐射能量总量或辐射度，即流 Φ 乘以传输光子的能量 E（$L = E\Phi$）。其中，$E = hc/\lambda$，与波长和普朗克常量 h 相关[Barrow 02]，且 c 表示为真空中的光速。

辐射度 L 表示为投影面积上单位实体角的、表面上的辐射流，如下所示。

$$L(r,\omega) = \frac{\mathrm{d}^2\Phi(r,\omega)}{\mathrm{d}^w \mathrm{d}S \cos\theta_s} \tag{16.6}$$

其中，θ_s 表示 w 和 $\mathrm{d}S$ 中表面法线间的夹角；因子 $\cos\theta_s$ 则表示基于立体角的能量流与表面角度无关。

辐射方程可通过下列公式给出：

$$L(r,\omega) = L_E(r,\omega) + \int_{\Theta_i} k(s,\omega' \to \omega)L(s,\omega')\cos\theta_s \mathrm{d}\omega' \tag{16.7}$$

上述方程也称作渲染方程，或者是真空中的单色辐射度时间不变方程，对应的几何描述如图 16.4（a）所示。

根据 r 中光照半球内的入射能量，式（16.7）的修正版本描述了源自点 r、方向为 ω_o 的能量。

$$L(r,\omega^o) = L_E(r,\omega^o) + \int_{\Theta_i} k(r,\omega \to \omega^o)L(s,\omega)\cos\theta_r \mathrm{d}\omega \tag{16.8}$$

其中，$\omega^o \in \Theta_o$，且 θ_r 表示为 r 中表面法线和 $\omega \in \Theta_i$ 间的夹角。图16.4（b）显示了该方程的几何描述，这种形式的辐射度方程可用于解释多种计算光照方法。

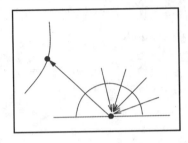 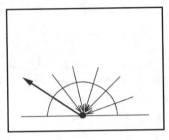

（a）式（16.7）的几何描述　　　（b）式（16.8）的几何描述

图16.4　式（16.7）和式（16.8）的几何描述

16.1.4　数值近似

光照方程可通过数值和近似法求解。这里将采用算子符号来求解积分方程。

相应地，积分算子 K 定义为：

$$(Kf)(x) = \int k(x,y) f(y) \mathrm{d}y$$

其中，函数 k 表示为算子的内核。这里，算子 K 通过 $(Kf)(x)$ 应用于函数 $f(x)$ 中。

当采用算子符号时，光照方程可写为下列形式：

$$L(r,\omega) = L_E(s,\omega) + (KL)(s,\omega) \tag{16.9}$$

或者

$$L = L_E + KL$$

在式（16.9）中，难点之一是函数 L 出现于等式的两侧（积分的内侧和外侧）。

针对这一问题，一种近似策略则采用了逐次代换法。需要注意的是，函数 L 通过一种自反（reflexive）的方式加以定义。最终，式（16.9）针对 L 提供了一个表达式。

具体来说，这里的基本思想是用等式右边的表达式替换 L，其中包括：

$$L = L_E + K(L_E + KL)$$
$$= L_E + KL_E + K^2 L$$

其中，指数表示算子 K 对函数 f 的连续应用，也就是说，$(K^2 f)(x) = (K(Kf))(x)$。重复这一次替换操作 $n+1$，则有：

$$L = L_E + K_E + \ldots + K^n L_E + K^{n+1} L$$
$$= \sum_{i=0}^{n} K^i L_E + K^{n+1} L$$

这种递归关系提供了一种近似计算 L 的方法。忽略 $n+1$ 阶的剩余项 $K^{n+1}L$，则可得到：

$$L \approx L_n = \sum_{i=0}^{n} K^i L_E$$

应用于光照函数计算的上述替换方法包含了较为直观的物理解释。注意，已知项 L_E 对应于光源发射的辐射能量。积分算子 K 对表面上的发射光线传播建模。因此，KL_E 对应于表面直接反射的光源光照，其连续应用模拟了反射光在场景中传播 n 次时的情形。

16.1.5 光照计算方法

前述内容讨论了光照函数的近似计算方法。在实际过程中，该策略将自身转换为两类光照计算方法，如下所示：

- 局部方法。$L_1 = L_E + KL_E$ 生成的近似结果对应于光源的直接贡献结果。此类方法仅使用了第 15 章介绍的局部光照模型。
- 全局方法。$L_n = \sum_{i=0}^{n} K^i L_E$ 生成的近似结果对应于光源的直接贡献结果，以及表面反射的间接贡献结果。此类方法使用了本章所讨论的全局光照。相应地，两种最为重要的实现形式分别是光线跟踪和辐射度方法。后续小节将对此予以解释。

16.2 光线跟踪方法

通过对场景中的光线路径进行采样，光线跟踪方法针对全局光照提供了一种解决方案，其基本理念可描述为，追踪场景中达到虚拟屏幕的现有光线。据此，该方法最恰当的名称应为反向光线跟踪。

光线跟踪机制适用于镜面反射（和传输）现象的建模，且依赖于虚拟相机。在该方法中，光照积分通过概率采样（使用蒙特卡罗积分）予以计算。稍后将会看到，对于完美镜面表面，该积分并无必要。

当通过光线跟踪方法求解光照方程时，可利用几何光学模型计算光子传输，旨在追踪加载辐射能量的粒子的路径。

此处将考查粒子 p，该粒子在场景中的路径对应于一系列的状态 $\{s_0, s_1, \cdots, s_n\}$。其间，在模拟的阶段 t 中，可将每种状态与 p 的属性相关联，如位置、方向和能量。

粒子通常包含一个存在区间，或称作生命周期。该时间间隔由与 p 相关的事件决定：它在初始状态 s_0 中生成，并在最终状态 s_n 时消失。光照方程描述了辐射能量的传输过程，即环境中光子（光线粒子）的传播。因此，在光线跟踪上下文中，可通过描述此类粒子状态变化的随机传输方程方便地制定该问题的公式，如下所示。

$$\ell(t) = g(t) + \int k(s \to t)\ell(s)\mathrm{d}s \qquad (16.10)$$

在该方程中，ℓ、g 和 k 可解释为概率分布。更准确地讲，$\ell(s_i)$ 表示状态 s_i 中可能存在的粒子数量；$g(s_i)$ 表示状态 s_i 中可能创建的粒子数量；$k(s_i \to s_j)$ 表示某个粒子跨越状态 s_i 和 s_j 间的概率。

当前需要计算场景表面上的光照函数。因此，可方便地将粒子状态与表面分解 $M = \cup M_i$ 相关的事件予以关联。因此可以说，当粒子路径到达表面 M_i 时，该粒子位于状态 s_i 中，如图 16.5 所示。

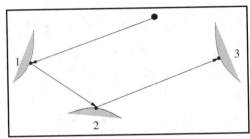

图 16.5　光线路径

当给定式（16.10）中的 g 和 k 后，可估算 ℓ。对此，可采用蒙特卡罗方法计算积分值。

注意，我们可以追踪沿路径前进或后退的粒子轨迹历史。在视见上下文环境中，可方便地从图像平面上的粒子开始，跟踪粒子创建位置（当光源发射粒子时）至该点间的路径，并对其予以记录。

离开状态 t_n 并返回至粒子轨迹历史中，则可得到：

$$\ell(t) \approx g(t_n) + \int k(s \to t_n)\ell(s)\mathrm{d}s$$

当前，$g(t_n)$ 为已知项，此处需要估算下列积分值：

$$\int k_n(s)\ell(s)\mathrm{d}s$$

其中，$k_n(s)$ 表示源自上一个状态 s、到达状态 t_n 时粒子的概率。

通过蒙特卡罗方法，可执行随机采样并估算 $\ell(t_{n-1})$，如下所示：

$$\begin{aligned}\ell(t_n) &\approx g(t_n) + \int k_n(s)\ell(s)\mathrm{d}s \\ &= g(t_n) + \ell(t_{n-1}) \\ &= g(t_n) + g(t_{n-1}) + \int k(t_{n-1} \to r)\ell(r)\mathrm{d}r\end{aligned}$$

持续上述处理过程,即可得到概率分布的近似估算结果 $\ell(t_n)$,如下所示:

$$\ell(t_n) \approx g(t_n) + g(t_{n-1}) + \ldots$$

需要注意的是,这一结果是根据之前推导的光照方程的计算方法得出的。

随机传输方程包含了与辐射度方程相同的结构,如下所示:

$$L(r,\omega_o) = L_E(r,\omega_o) + \int_\Theta k(r,\omega \to \omega^o) L(s,\omega) \cos\theta_r \, d\omega$$

当采用概率方法求解上述方程时,可通过两种技术高效地估算基于蒙特卡罗方法的积分值:

- 分层。层中的光照半球 $\Theta = \bigcup \Pi_m$ 的分解如下所示:

$$D_m = \{\omega; \omega \in \Pi_m\}$$

以便函数 $L(S,\omega)$,$\omega \in \Pi_m$ 呈现较小的变化。

- 重要性采样。在每层中的对 $L(s,\omega)$ 进行采样,并选择更具代表性的样本。这可通过重要性函数予以实现,如下所示:

$$g_m : R^3 \times S^2 \to [0, 1]$$

这里,通过将上述两种技术整合至辐射度方程中,可得到:

$$L(r,\omega_o) = L_E(r,\omega_o) + \sum_{m=1}^{M} \int_{D_m} k(r,\omega \to \omega^o) \frac{L(s,\omega)\cos\theta_r}{g_m(r,\omega)} g_m(r,\omega) d\omega$$

其中,各层 D_i 通过方向集合 $\omega \in \Pi_i$ 给出。注意,此处将 L 除以 g_m,并于随后将该结果乘以 g_m,以避免向方程中引入误差。

通过可见性函数 $N(r,\omega) = r - v(r,\omega)\omega$,我们得到了每层中的可见表面点。通过这种方式,可确定每层 D_i 的贡献,如下所示:

$$\int_{D_i} = \int_{w \in \Pi_i} k(r,\omega \to \omega^o) L(N(r,\omega),\omega) \cos\theta \, d\omega$$

简而言之,计算光照的光线跟踪方法的通用模式包含以下步骤:

(1)利用 $\Theta = \bigcup \Pi_m$ 选择层 $\{D_m\}$。
(2)确定层 D_m 的可见性 $N(s,\omega)$。
(3)在各层 D_m 中估算光照积分。

分层和重要性函数应基于局部表面信息和全局场景信息。一种较好的选择方案是将此类信息分为两层,如下所示:

- 直接光照。其中,层次根据光源信息计算。
- 间接光照。层次根据表面的双向反射函数计算。

经典的光线跟踪算法使用了两个假设条件,并对当前问题予以简化,即点光源和完美镜面表面。通过这种方式,反射函数对应于 Dirac delta 分布[Strichartz 94],分层则归结于方向的离散集合。随后，光照方程可简化为：

$$L(r,\omega) = \sum_{w'}[k_l(r,\omega \to \omega')L_E(s_l,\omega')] + [k_r L(s_r,\omega^r) + k_t L(s_t,\omega^t)] \quad (16.11)$$

其中，$s_i = N(r,\omega^i) = r - v(r,\omega^i)\omega^i$，且 $i = l$、s、t，由沿光源方向的光线 ω^l、ω^r 的反射光线和 ω^t 的折射光线给出。需要注意的是，方程的第一部分对应于光源的直接光照；而第二部分则对应于镜面间接光照。另外，第二部分内容采用递归方式计算。

接下来将展示该算法的实现过程。

例程 ray_shade 通过光线跟踪技术计算光照。

```
275 <ray shade 275>≡
  Color ray_shade(int level, Real w, Ray v, RContext *rc, Object *ol)
  {
    Inode *i = ray_intersect(ol, v);
    if (i != NULL) { Light *l; Real wf;
      Material *m = i->m;
      Vector3 p = ray_point(v, i->t) ;
      Cone recv = cone_make(p, i->n, PIOVER2);
      Color c = c_mult(m->c, c_scale(m->ka, ambient(rc)));

      for (l = rc->l; l != NULL; l = l->next)
        if ((*l->transport)(l, recv, rc) && (wf = shadow(l, p, ol))
           > RAY_WF_MIN)
          c = c_add(c, c_mult(m->c,
                c_scale(wf * m->kd * v3_dot(l->outdir,i->n), l->outcol)));

      if (level++ < MAX_RAY_LEVEL) {
        if ((wf = w * m->ks) > RAY_WF_MIN) {
          Ray r = ray_make(p, reflect_dir(v.d, i->n));
          c = c_add(c, c_mult(m->s,
             c_scale(m->ks, ray_shade(level, wf, r, rc, ol))));
        }
        if ((wf = w * m->kt) > RAY_WF_MIN) {
          Ray t = ray_make(p, refract_dir(v.d, i->n, (i->enter)?
                                    1/m->ir: m->ir));
          if (v3_sqrnorm(t.d) > 0) {
            c = c_add(c, c_mult(m->s,
                c_scale(m->kt, ray_shade(level, wf, t, rc, ol))));
          }
        }
      }
    }
  }
```

第16章 全局光照

```
      inode_free(i);
      return c;
    } else {
      return BG_COLOR;
    }
  }
Defines:
  ray_shade, never used.
Uses BG_COLOR, MAX_RAY_LEVEL, RAY_WF_MIN, reflect_dir 276, refract_dir
  277a, and shadow 277b.
```

当光线路径超出 MAX_RAY_LEVEL，或者剩余路径的贡献估算结果 wf 小于 MAX_WF_MIN 时，递归过程将终止。

例程 reflect_dir 计算反射光线的方向。

```
276 <reflect dir 276>≡
  Vector3 reflect_dir(Vector3 d, Vector3 n)
  {
    return v3_add(d, v3_scale(-2 * v3_dot(n, d), n));
  }
Defines:
  reflect_dir, used in chunk 275.
```

例程 refract_dir 利用斯涅耳定律计算透射光线的方向。

```
277a <refractdir 277a>≡
  Vector3 refract_dir(Vector3 d, Vector3 n, Real eta)
  {
    Real c1, c2;

    if ((c1 = v3_dot(d, n)) < 0)
      c1 = -c1;
    else
      n = v3_scale(-1.0, n);

    if ((c2 = 1 - SQR(eta) * (1 - SQR(c1))) < 0)
      return v3_make(0,0,0);
    else
      return v3_add(v3_scale(eta, d), v3_scale(eta*c1 - sqrt(c2), n));
  }
Defines:
refract_dir, used in chunk 275.
```

例程 shadow 判断光源方向上的光线是否位于阴影区域内。

```
277b <shadow 277b>≡
  Real shadow(Light *l, Vector3 p, Object *ol)
  {
    Real t, kt; Inode *i; Vector3 d;

    if (l->type ! = LIGHT_POINT)
      return 1.0;
    if ((i = ray_intersect(ol, ray_make(p, d))) == NULL)
      return 1.0;
    t = i->t; kt = i->m->kt; inode_free(i);

    if (t > RAY_EPS && t < 1)
      return 0.0;
    else
      return 1.0;
  }

Defines:
  shadow, used in chunk 275.
```

图 16.6 显示了光线跟踪算法中所采用的模式。

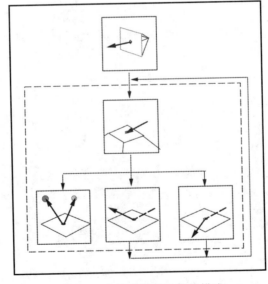

图 16.6　光线跟踪算法中的模式

16.3 辐射度方法

根据场景表面的离散结果,辐射度方法针对全局光照计算提供了一种解决方案,其基本思想可描述为:分解多边形元素中的表面,并计算所有此类元素的能量交换。针对于此,辐射度方法特别适用于漫反射交互建模,且与虚拟相机无关。

在辐射度方案中,光照积分通过基于 Galerkin 方法的有限元进行计算。式(16.8)描述了点 r 沿方向 ω^o 传播的辐射度 $L(r,\omega^o)$,并作为该点处的发射和入射能量函数,如下所示:

$$L(r,\omega^o) = L_E(r,\omega^o) + \int_{\Theta_i} k(r,\omega \to \omega^o) L(s,\omega) \cos\theta_r \, d\omega \quad (16.12)$$

利用可见性函数 v,此处定义了沿方向 ω 到达 r 处的能量传输,以使 $s = r + v(r,s)\omega$。若 $s \in dS$ 位于一个较远的表面上,实体角 $d\omega$ 可记为:

$$d\omega = \frac{dS \cos\theta_s}{\|r-s\|^2} \quad (16.13)$$

其中,θ_s 表示为 dS 法线和向量 $(r-s)$ 间的夹角。

在将上述表达式代入光照方程中时,因确保积分仅对可见表面上的点执行。因此,下面定义了可见函数测试:

$$V(r,s) = \begin{cases} 1, & \text{若 } s = r - v(r,s-r)(s-r); \\ 0, & \text{其他} \end{cases} \quad (16.14)$$

将方程中立体角的表达式代入,同时引入可见性函数,即可通过可见表面上的面积修改光照半球中实体角的积分域。通过这种方式,可得到:

$$L(r,\omega^o) = L_E(r,\omega^o) + \int_M k(r,\omega \to \omega^o) L(s,\omega) G(r,s) \, d\omega \quad (16.15)$$

其中,函数

$$G(r,s) = \frac{\cos\theta_s \cos\theta_r}{\|r-s\|^2} V(r,s) \quad (16.16)$$

仅取决于几何形状,如图 16.7 所示。

该离散方法通过多边形面片 $M_i = \cup m_k$(有限元)划分表面,并定义了函数基 $\{b_j\}_{j \in J}$,并生成了场景表面上的近似空间。该空间内的解 $L(r,\omega)$ 的投影可记为基函数的线性组合,如下所示:

$$\hat{L}(r,\omega) = \sum_j L_j b_j(r,\omega) \quad (16.17)$$

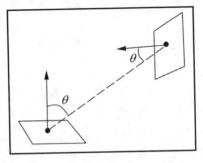

图 16.7 点到点传输的几何描述

计算该函数空间内的方程投影,可得到:

$$\langle \hat{L}, b_i \rangle = \langle L_E, b_i \rangle + \left\langle \int_M k(r,\omega)G(r,s)\hat{L}, b_i \right\rangle \quad (16.18)$$

替换方程中的 $\hat{L}$ 表达式,可得到:

$$\left\langle \sum_j L_j b_j, b_i \right\rangle = \langle L_E, b_i \rangle + \left\langle \int_M k(r,\omega)G(r,s)\sum_j L_j b_j, b_i \right\rangle \quad (16.19)$$

重新排列 L_j 中的各项,并移除内积的和,可得到:

$$\langle L_E, b_i \rangle = \left\langle \sum_j L_j b_j, b_i \right\rangle - \left\langle \int_M k(r,\omega)G(r,s)\sum_j L_j b_j, b_i \right\rangle$$

$$\langle L_E, b_i \rangle = \sum_j L_j \left[\langle b_j, b_i \rangle - \left\langle \int_M k(r,\omega)G(r,s)b_j, b_i \right\rangle \right]$$

注意,可将上述表达式记为矩阵 $L_E = KL$ 形式。

经典的辐射度方法设置了以下假设条件,进而简化当前任务:

(1) 表面为不透明。不存在透射传播。
(2) Lambertian 反射。存在完全漫反射表面。
(3) 辐射度和辐照度在各个元素内保持恒定。

漫反射表明,双向反射函数 $k(s, \omega \to \omega')$ 在各个方向上保持恒定,因而与 ω 无关。最终,可利用积分外的函数 $\rho(s)$ 对其进行替换,如下所示:

$$\int_M k(s, \omega \to \omega')L(s,\omega)G(r,s)\mathrm{d}\omega = \rho(s)\int_M L(s,\omega)G(r,s)\mathrm{d}\omega \quad (16.20)$$

据此,还可生成替换操作,并将辐射转换为辐射度: $L\pi = B$。

分段式恒定光照函数表明,可对有限元近似空间采用 Haar 基 $\{b_i\}$。

$$b_i(r) \begin{cases} 1, & r \in M_i \\ 0, & \text{其他} \end{cases} \quad (16.21)$$

由于 Haar 基函数是非连续的,因而有:

$$\langle b_i, b_j \rangle = \delta_{ij} A_i \tag{16.22}$$

整合上述数据，Haar 基中的光照积分变为：

$$\left\langle \int_M k(r,\omega)\, G(r,s) b_j, b_i \right\rangle = \frac{\rho_i}{\pi} \int_{M_i} \int_{M_k} G(i,k) \mathrm{d}k \mathrm{d}i = \rho_i A_i F_{i,k} \tag{16.23}$$

其中

$$F_{i,k} = \frac{1}{A_i} \int_{M_i} \int_{M_k} \frac{\cos\theta_i \cos\theta_k}{\pi \| i-k \|^2} V(i,k) \mathrm{d}k \mathrm{d}i \tag{16.24}$$

表示为形状因子，表示离开元素 i、到达元素 j 的辐射能量百分比。

利用下列公式：

$$\langle L_E, b_i \rangle = \int_{M_i} L_E(s) \mathrm{d}s = E_i A_i$$

并将 $L \mapsto \dfrac{B}{\pi}$ 代入方程中，可得到：

$$E_i A_i = \sum_k B_k (\delta_{ik} A_i - \rho_i A_i F_{i,k})$$

$$E_i A_i = B_i A_i - \rho_i \sum_k B_k A_i F_{i,k}$$

或者，等式两边除以 A_i 并重新排列各项，可得到：

$$B_i = E_i + \rho_i \sum_k B_k F_{i,k} \tag{16.25}$$

上述方程称作经典辐射度方程。在实际操作过程中，将针对离散的 n 元素的辐射度 B 设置一个包含 n 个方程的方程组，其矩阵形式如下所示：

$$B = E + FB$$
$$(I - F)B = E$$

或者

$$\begin{pmatrix} 1-\rho_1 F_{11} & \cdots & -\rho_1 F_{1n} \\ -\rho_2 F_{21} & \cdots & -\rho_2 F_{2n} \\ \cdots & & \\ -\rho_n F_{n1} & \cdots & 1-\rho_n F_{nn} \end{pmatrix} \begin{pmatrix} B_1 \\ B_2 \\ \cdots \\ B_n \end{pmatrix} = \begin{pmatrix} E_1 \\ E_2 \\ \cdots \\ E_n \end{pmatrix}$$

鉴于有限元内的表面离散化通过多边形网格实现，因而有 $F_{ii} = 0$，且矩阵的对角线为 1。

此处需要通过下列方式计算方程组的数值解：

$$B = (I-F)^{-1} E$$

当线性方程组较大时，逆矩阵则变得不切实际。此时，较好的方法是根据逐步求精原则采用迭代方法。其中，当给定线性方程组 $Mx = y$，可生成一系列的近似解 x^k，当

$k \to \infty$ 时，对应解将收敛于 x。

阶段 k 中的近似误差可通过下列公式给出：
$$e^k = x - x^k \tag{16.26}$$

$Mx^k = y + r^k$ 近似计算导致的剩余项 r^k 表示为：
$$r^k = y - Mx^k \tag{16.27}$$

当最小化每个阶段 k 中的剩余项 r^k 时，可根据误差表示剩余项。此处减去 r^k 的 $y - Mx = 0$，如下所示：
$$\begin{aligned} r^k &= (y - Mx^k) - (y - Mx) \\ &= M(x - x^k) \\ &= Me^k \end{aligned}$$

迭代方法的基本思想是，改进近似结果 x^k，进而生成较优的近似结果 x^{k+1}。此处将采用一种称之为 Southwell 的求解方法，该方法将获取某个转换，以使下一阶段的元素 x_i^{k+1} 之一的剩余项 r_i^{k+1} 等于 0。通过这种方式，可利用较大的剩余项选择元素 x^i，并计算 x_i^{k+1}，进而满足下列条件：
$$r_i^{k+1} = 0$$
$$y_i + \sum_j M_{ij} x_j^{k+1} = 0$$

仅当向量 x 的分量 i 变化时，可得到 $x_j^{k+1} = x_j^k$，且 $j \neq i$。x_i^{k+1} 的新值如下所示：
$$\begin{aligned} x_i^{k+1} &= \frac{1}{M_{ii}}\left(y_i - \sum_{i \neq j} M_{ij} x_j^k\right) \\ &= x_i^k + \frac{r_i^k}{M_{ii}} \\ &= x_i^k + \Delta x_i^k \end{aligned}$$

新的剩余项可通过下列方式计算：
$$\begin{aligned} r^{k+1} &= y - Mx^{k+1} \\ &= y - M(x^k + \Delta x^k) \\ &= y - Mx^k - M\Delta x^k \\ &= r^k - M\Delta x^k \end{aligned}$$

然而，除了 Δx_i^k 之外，向量 $\Delta x^k = x^{k+1} - x^k$ 包含了全部等于 0 的分量，因而有：
$$r_j^{k+1} = r_j^k - \frac{K_{ji}}{K_{ii}} r_i^k \tag{16.28}$$

需要注意的是，当更新剩余项的向量时，仅使用了矩阵的列，如下所示。

$$\begin{pmatrix} x \\ x \\ x \\ x \end{pmatrix} = \begin{pmatrix} x \\ x \\ x \\ x \end{pmatrix} + \begin{pmatrix} . & x & . & . \\ x & . & . & . \\ . & x & . & . \\ . & x & . & . \end{pmatrix}$$

改进后的辐射度算法使用了 Southwell 方法的变化版本，并利用较少的迭代生成了较好的近似解。

在光照问题环境中，可将剩余项 R_i^k 解释为：未在场景中传播的、阶段 k 中元素 M_i 的辐射能量。

根据这一物理解释，可以看出，Southwell 方法通过其他元素 M_j ($j \neq i$) 更新元素 M_i 的非散射辐射能量。另外一方面，此类元素还将散射接收自后续阶段的能量。

此时，$M = (I - F)$。鉴于已知 $M_{ii} = 1$，且 $M_{ij} = -\rho_j F_{ij}$，因而更新剩余项向量时可得到以下规则：

$$R_j^{k+1} = R_j^k + (\rho_j F_{ij}) R_i^k$$

且有 $R_i^{k+1} = 0$。在改进后的辐射度中，还可进一步更新元素 B_j ($j \neq i$) 的辐射向量，如下所示：

$$B_j^{k+1} = B_j^k + (\rho_j F_{ij}) R_i^k$$

例程 radiosity_prog 实现了改进后的辐射度算法。

```
283 <radiosity prog 283>≡
 Color *radiosity_prog(int n, Poly **p, Color *e, Color *rho)
 {
   int src, rcv, iter = 0;
   Real ff, mts, *a = NEWTARRAY(n, Real);
   Color d, *dm = NEWTARRAY(n, Color);
   Color ma, *m = NEWTARRAY(n, Color);

   initialize(n, m, dm, a, p, e);
   while (iter-- < max_iter) {
     src = select_shooter(n, dm, a);
     if (converged(src, dm))
       break;
     for (rcv = 0; rcv < n; rcv++) {
       if (rcv == src|| (ff = formfactor(src, rcv, n, p, a)) < REL_EPS)
         continue;
       d = c_scale(ff, c_mult(rho[rcv], dm[src] ));

       m[rcv] = c_add(m[rcv], d);
```

```
      dm[rcv] = c_add(dm[rcv], d);
    }
    dm[src] = c_make(0,0,0);
  }
  ma = ambient_rad(n, dm, a);
  for (rcv = 0; rcv < n; rcv++)
    m[rcv] = c_add(m[rcv], ma);
  efree(a), efree(dm);
  return m;
}
Defines:
  radiosity_prog, never used.
Uses ambient_rad 286, converged 285c, formfactor 285a, initialize 284a,
max_iter, and select_shooter 284b.
```

例程 initialize 初始化数值属性。

```
284a <init radiosity 284a>≡
  static void initialize(int n, Color *m, Color *dm, Real *a, Poly **p,
                         Color *e)
  {
    int i;
    for (i = 0; i < n; i++) {
      a[i] = poly3_area(p[i] );
      m[i] = dm[i] = e[i] ;
    }
  }
Defines:
  initialize, used in chunk 283.
```

例程 select_shooter 利用较大非散射辐射选择元素。

```
284b <select shooter 284b>≡
  static int select_shooter(int n, Color *dm, Real *a)
  {
    Real m, mmax; int i, imax;

    for (i = 0; i < n; i++) {
      m = c_sqrnorm(c_scale(a[i], dm[i] ));
      if (i == 0|| m > mmax) {
        mmax = m; imax = i;
      }
    }
    return imax;
  }
```

Defines:
  select_shooter, used in chunk 283.

例程 formfactor 计算形状因子 F_{ij}。

285a <form factor 285a>≡
```
static Real formfactor(int i, int j, int n, Poly **p, Real *a)
{
  Real r2, ci, cj; Vector3 vi, vj, vji, d;
  vi = poly_centr(p[i] );
  vj = poly_centr(p[j] );
  vji = v3_sub(vi, vj);
  if ((r2 = v3_sqrnorm(vji)) < REL_EPS)
    return 0;
  d = v3_scale(1.0/sqrt(r2), vji);
  if ((cj = v3_dot(poly_normal(p[j] ), d)) < REL_EPS)
    return 0;
  if ((ci = -v3_dot(poly_normal(p[i] ), d)) < REL_EPS)
    return 0;
  if (vis_flag && visible(n, p, vj, vji) < REL_EPS)
    return 0;
  return a[i] * ((cj * ci) / (PI * r2 + a[i] ));
}
```
Defines:
  formfactor, used in chunk 283.
Uses vis_flag and visible 285b.

例程 visible 判断两个元素 i、j 之间的可见性。

285b <visible 285b>≡
```
static Real visible(int n, Poly **p, Vector3 v, Vector3 d)
{
  Ray r = ray_make(v, d);

  while (n--) {
    Real t = poly3_ray_inter(p[n], poly3_plane(p[n] ), r);
    if (t > REL_EPS && t < 1)
      return 0.0;
  }
  return 1.0;
}
```
Defines:
  visible, used in chunk 285a.

例程 converged 测试解的收敛性。

```
285c <converged 285c>≡
  static int converged(int i, Color *dm)
  {
    return (c_sqrnorm(dm[i] ) < dm_eps);
  }
Defines:
  converged, used in chunk 283.
Uses dm_eps.
```

例程 ambient_rad 计算环境辐射度。

```
286 <ambientrad 286>≡
  static Color ambient_rad(int n, Color *dm, Real *a)
  {
    int i; Real aa = 0;
    Color ma = c_make(0,0,0);
    for (i = 0; i < n; i++) {
      ma = c_add(ma, c_scale(a[i], dm[i] ));
      aa += a[i] ;
    }
    return c_scale(1.0/aa, ma);
  }
Defines:
  ambient_rad, used in chunk 283.
```

图 16.8 显示了改进后的算法的处理模式。

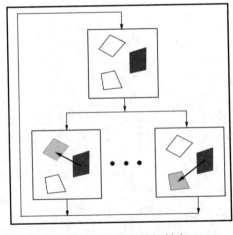

图 16.8　改进后的辐射度

16.4 补充材料

图 16.9 显示了通过光线跟踪计算后的场景。图 16.10 则显示了基于辐射度的光照计算程序截图。

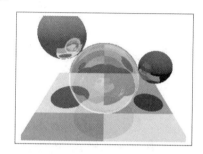

图 16.9　光线跟踪

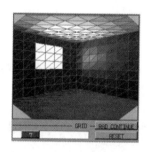

图 16.10　辐射度

GLOBAL 库的 API 包含以下例程：

```
Color ray_shade(int level, Ray v, Inode *i, RContext *rc);
Ray ray_reflect(Ray r, Inode *i);
Ray ray_refract(Ray r, Inode *i);

Inode *ray_intersect(Ray r, RContext *rc);
Color *progress_rad(int n, Poly **p, Color *e, Color *rho, Real eps);

void initialize(int n, Color *m, Color *dm, Real * a, Poly **p, Color *e);
int converged(int n, Color *dm, Real *a, Real eps);
int select_shooter(int n, Color *dm, Real *a);
Real formfactor(int i, int j, int n, Poly **p, Real *a);
Real visible(int n, Poly **p, Vector3 v, Vector3 d);
Color ambient_rad(int n, Color *dm, Real *a);
```

16.5 本章练习

（1）生成一个简单的场景，其中包含两个球体和一个光源。编写一个程序并利用光线跟踪机制计算光照。

（2）生成一个简单的场景，其中包含一个封闭的盒体。将光源与盒体的某个面关联。编写一个程序，并利用辐射度计算光照。

第 17 章 贴图技术

本章将讨论贴图技术,并定义图形对象的属性函数。

17.1 基础知识

当定义属性函数时,贴图机制是一种功能强大的技术。在本节中将会看到,全部贴图技术均是单一数学模型中的一部分内容,该模型可用于研究视见处理过程中的多偶制纹理应用程序。

17.1.1 贴图的概念

纹理可视为欧几里得空间$\mathbb{R}^m$的子集到欧几里得空间$\mathbb{R}^k$的一种应用,即$t: U \subset \mathbb{R}^m \to \mathbb{R}^k$。名称"纹理"源自$m=2$,$k=3$这一特殊情形;同时,欧几里得空间$\mathbb{R}^k$通过颜色空间进行标识,其中,$t$代表数字图像。

当给定对象空间子集$V \subset \mathbb{R}^n$的函数$g: V \to U \subset \mathbb{R}^m$,我们称纹理贴图为应用程序的组成,如下所示:

$$\tau = g \circ t : V \to \mathbb{R}^k$$

并将向量空间$\mathbb{R}^k$的一个元素$t(g(x,y,z))$与对象空间的各点(x,y,z)关联。

这一模式如图 17.1 所示。

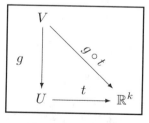

图 17.1 纹理贴图

纹理贴图涉及以下空间。

❑ 对象空间:$V \subset \mathbb{R}^n$。

- 纹理空间：$U \subset \mathbb{R}^m$。
- 属性空间：$\mathbb{R}^k$。

上述空间与下列函数相关。

- 纹理函数：$t: U \subset \mathbb{R}^m \to \mathbb{R}^k$。
- 贴图函数：$g: V \subset \mathbb{R}^n \to U$。

纹理函数负责构建几何纹理域及其属性值之间的关联。贴图函数则构建对象点和纹理域各点间的对应关系。

17.1.2 贴图类型

纹理函数取决于纹理空间 $\mathbb{R}^m$ 的维度 m 和属性空间 $\mathbb{R}^k$ 的性质。根据纹理域的维度，可包含 3 种纹理贴图类型，如下所示。

- 1D 贴图。当 $m=1$ 时，t 表示为颜色贴图。
- 2D 贴图。当 $m=2$ 时，t 表示为数字图像。
- 3D 贴图。当 $m=3$ 时，t 称作实体纹理。

贴图函数取决于图形对象支撑 V 的几何描述，以及纹理空间的维度。对此，存在两种基本方式可定义纹理函数。第一种方法基于与对象关联的参数化过程；第二种方法则基于某种投影类型。

当纹理空间包含维度 2，且 V 表示为一个参数表面时，参数化提供了一种更为自然的方法。此时，可得到一个表面方程 $f: \mathbb{R}^2 \to V$，其反函数 f^{-1} 可视为贴图函数 g。

当纹理空间包含维度 2，且 V 定义为隐式表面时，则推荐使用投影方法，并于其中寻找一种适用于对象形状的投影。对此，较为常用的是球体和圆柱投影。

当纹理空间包含维度 3 时，可将 $\mathbb{R}^3$ 到 $U \subset \mathbb{R}^3$ 点间的约束条件作为贴图函数。该方案同样适用于采用参数和隐式方式定义的表面。

17.1.3 贴图应用

纹理贴图应用与属性紧密相关，其中较为重要的应用包括：

- 外观，即对象的表面属性，如颜色、反射率等。
- 几何形状，用于对象的几何建模，例如，指定表面上的粗糙度。
- 光照，表示为光照计算中的数据，其中，贴图机制可作为视见算法的缓存引擎。相关示例包括贴图反射和阴影。

17.2 节将讨论上述贴图应用的实现过程。

17.2 纹理函数

本节将介绍二维纹理函数的实现，其中涉及通用表达形式和两种定义类型，即采样和过程式定义。

17.2.1 表达方式

通用 2D 纹理通过数据结构 TextureSrc 予以表示，其中包含了一个访问函数和纹理数据。

```
291a <texture source 291a>≡
  typedef struct TextureSrc {
    Color (*texfunc)();
    void *texdata;
  } TextureSrc;
Defines:
  TextureSrc, used in chunks 291b, 292b, 294b, 297b, and 299.
```

纹理可通过一幅图像（TEX_IMG 类型）或一个函数（TEX_FUNC 类型）描述。例程 parse_texsource 执行场景描述语言中的通用纹理封装操作。

```
291b <parse texture source 291b>≡
  TextureSrc *parse_texsource(Pval *pl)
  {
    Pval *p = pl;
    while (p != NULL) {
      if (p->name && strcmp(p->name,"tex_src") == 0) {
        if (p->val,type == TEX_IMG|| p->val,type == TEX_FUNC)
          return (TextureSrc *)(p->val,u,v);
      }
      p = p->next;
    }
    return default_tsrc();
  }
Defines:
  parse_texsource, used in chunks 295b and 301.
Uses default_tsrc and TextureSrc 291a.
```

17.2.2 图像定义

当纹理通过一幅图像定义时，结构 TextureSrc 中的 texdata 字段表示为 Image 类型。

对应的访问函数为 image_texture 例程。

```
292a <image texture 292a>≡
  Color image_texture(Image *i, Vector3 t)
  {
    Color c00, c01, c10, c11, c;
    Real ru, rv, tu, tv; int u, v;
    ru = t.x * i->w;
    rv = (1 - t.y) * i->h;
    u = floor(ru); tu = ru - u;
    v = floor(rv); tv = rv - v;
    c00 = img_getc(i, u, v);
    c01 = img_getc(i, u, v+1);
    c10 = img_getc(i, u+1, v);
    c11 = img_getc(i, u+1, v+1);
    c = v3_bilerp(tu, tv, c00, c01, tv, c10, c11);
    return c_scale(1./255.);
  }
Defines:
  image_texture, used in chunk 292b.
```

例程 imagemap_parse 实现了通过图像（存储为光栅化格式）定义的纹理贴图。

```
292b <parse image map 292b>≡
  Val imagemap_parse(int pass, Pval *pl)
  {
    Val v;
    if (pass == T_EXEC) {
      Pval *p = pl;
      TextureSrc *i = NEWSTRUCT(TextureSrc);
      i->texfunc = texture_default;
      while (p != NULL) {
        if (p->name && strcmp(p->name,"fname") == 0 && p->val.type == V_STR) {
          i->texfunc = image_texture;
          i->texdata = img_read(p->val,u,v);
        }
        p = p->next;
      }
      v.type = TEX_IMG;
      v.u.v = i;
    }
    return v;
  }
```

```
Defines:
  imagemap_parse, never used.
Uses image_texture 292a, texture_default, and TextureSrc 291a.
```

例程 imagemap_parse 与下列场景描述语言的构建相关联:

```
tex_src = imagemap { fname = "name.ras"}}
```

17.2.3 过程式定义

当纹理采用过程方式加以定义时,访问函数将利用存储于 texdata 字段中的参数生成属性值。

下面将生成一个棋盘模式的过程式纹理。例程 chequer_texture 实现了一个棋盘纹理,并在例程 API 中定义了其参数。

```
293a <chequer texture 293a>≡
  Color chequer_texture(ChequerInfo *c, Vector3 t)
  {
    return chequer(c->xfreq, c->yfreq, c->fg, c->bg, t);
  }
Defines:
  chequer_texture, never used.
Uses chequer 293b and ChequerInfo.
```

例程 chequer 计算棋盘模式的颜色值。

```
293b <chequer 293b>≡
  Color chequer(Real freq, Color a, Color b, Vector3 t)
  {
    Real sm = mod(t,x * freq, 1);
    Real tm = mod(t,y * freq, 1);

    return ((sm < ,5) ? ((tm < ,5)? a : b) : ((tm < ,5)? b : a));
  }
Defines:
  chequer, used in chunk 293a.
Uses mod 294a.
```

例程 mod 用作辅助函数,并生成棋盘纹理。

```
294a <mod 294a>≡
  Real mod(Real a, Real b)
  {
```

```
    int n = (int)(a/b);
    a -= n * b;
    if (a < 0)
      a += b;
    return a;
  }
Defines:
 mod, used in chunks 293b and 294c.
```

17.3 纹理贴图

纹理贴图将属性值 $t(g(x, y, z))$ 与对象支撑 V 的各点 (x, y, z) 关联。这里，属性值用于计算给定点处的光照函数。对应属性可以是表面的颜色值，或者是对象材质的其他属性。当纹理空间包含维度 2 时，对应结果为对象表面上处于拉伸状态的弹性环绕效果。

例程 TmapInfo 存储了纹理贴图数据。

```
294b <tmap info 294b>≡
  typedef struct TmapInfo {
    TextureSrc *src;
    int code;
    Color bg;
  } TmapInfo;
Defines:
 TmapInfo, used in chunks 294c and 295b.
Uses TextureSrc 291a.
```

例程 texture_map 执行纹理贴图操作。其中，纹理应用程序可以是 TMAP_TILE 类型或 TMAP_CLAMP 类型。

```
294c <texture map 294c>≡
  Color texture_map(void *info, Vector3 t)
  {
    TmapInfo *i = info;

    switch (i->code) {
    case TMAP_TILE:
      t,x = mod(t,x, 1); t,y = mod(t,y, 1);
      break;
    case TMAP_CLAMP:
      if (t,x < 0| | t,x > 1| | t,y < 0| | t,y > 1)
```

第 17 章 贴图技术

```
      return i->bg;
    break;
  }
  return (*i->src->texfunc)(i->src->texdata, t);
}
```
Defines:
  texture_map, used in chunk 295a.
Uses mod 294a, TMAP_CLAMP, TMAP_TILE, and TmapInfo 294b.

例程 textured_plastic 实现了塑料材质，其颜色用纹理给出。

```
295a <textured plastic 295a>≡
  Color textured_plastic(RContext *rc)
  {
    Color ct = texture_map(rc->m->tinfo, rc->t);

    return c_add(c_mult(ct, c_add(c_scale(rc->m->ka, ambient(rc)),
                                  c_scale(rc->m->kd, diffuse(rc)))),
                 c_mult(rc->m->s, c_scale(rc->m->ks, specular(rc))));
  }
```
Defines:
  textured_plastic, used in chunk 295b.
Uses texture_map 294c.

例程 textured_parse 定义了场景描述语言中的塑料纹理材质。

```
295b <parse textured 295b>≡
  Val textured_parse(int pass, Pval *pl)
  {
    Val v;
    if (pass == T_EXEC) {
      Material *m = NEWSTRUCT(Material);
      TmapInfo *ti = NEWSTRUCT(TmapInfo);
      ti->src = parse_texsource(pl);
      ti->bg = pvl_get_v3(pl, "bg_col", C_WHITE);
      ti->code = parse_code(pl);
      m->tinfo = ti; m->luminance = textured_plastic;
      v.u.v = m; v.type = MATERIAL;
    }
    return v;
  }
```
Defines:
  textured_parse, never used.

```
Uses parse_code 296, parse_texsource 291b, textured_plastic 295a, and
TmapInfo 294b.
```

例程 parse_code 针对表面纹理解释了应用程序类型的代码。

```
296 <parse code 296>≡
  int parse_code(Pval *pl)
  {
    Pval *p = pl;
    while (p ! = NULL) {
      if (p->name && strcmp(p->name,"code") == 0 && p->val,type) {
        if (strcmp(p->val,u,v, "tile") == 0)
          return TMAP_TILE;
        else if (strcmp(p->val,u,v, "clamp") == 0)
          return TMAP_CLAMP;
      }
      p = p->next;
    }
    return TMAP_CLAMP;
  }
Defines:
  parse_code, used in chunk 295b.
Uses TMAP_CLAMP and TMAP_TILE.
```

图 17.2 显示了纹理贴图示例。其中一个球体使用了过程式纹理，而另一个球体则使用了一幅图像。

图 17.2　纹理贴图

17.4　凹凸贴图

粗糙度贴图也称作凹凸纹理，并执行了表面法线的扰动行为，进而模拟几何体表面的不规则现象。

例程 rough_surface 利用凹凸贴图实现了不规则表面材质。

```
297a <rough surface 297a>≡
  Color rough_surface(RContext *rc)
  {
    Vector3 d = bump_map(rc->m->tinfo, rc->t, rc->n, rc->du, rc->dv);
    rc->n = v3_unit(v3_add(rc->n, d));
    return matte(rc);
  }
Defines:
  rough_surface, never used.
Uses bump_map 297b.
```

例程 bump_map 计算应用于法线向量的扰动行为。

```
297b <bump map 297b>≡
  Color bump_map(void *info, Vector3 t, Vector3 n, Vector3 ds, Vector3 dt)
  {
    TextureSrc *src = info;
    Real h = 0.0005;
    Real fo = texture_c1((*src->texfunc)(src->texdata, t));
    Real fu = texture_c1((*src->texfunc)(src->texdata,
                                          v3_add(t,v3_make(h,0,0))));
    Real fv = texture_c1((*src->texfunc)(src->texdata,
                                          v3_add(t,v3_make(0,h,0))));
    Real du = fderiv(fo, fu, h);
    Real dv = fderiv(fo, fv, h);
    Vector3 u = v3_scale(du, v3_cross(n, dt));
    Vector3 v = v3_scale(-dv, v3_cross(n, ds));

    return v3_add(u, v);
  }
Defines:
  bump_map, used in chunk 297a.
Uses fderiv 298a, texture_c1 297c, and TextureSrc 291a.
```

例程 texture_c1 返回颜色向量中的第一个分量。

```
297c <texture c1 297c>≡
  Real texture_c1(Color c)
  {
    return c.x;
```

```
        }
Defines:
  texture_c1, used in chunk 297b.
```

例程 fderiv 通过有限差分计算导数。

```
298a <fderiv 298a>≡
  Real fderiv(Real f0, Real f1, Real h)
  {
    return (f1 - f0)/h;
  }
Defines:
  fderiv, used in chunk 297b.
```

图 17.3 显示了一个凹凸贴图示例。

图 17.3　凹凸贴图

17.5　反射贴图

在反射贴图中，对象将反射纹理 t。对象各点(x, y, z)的投影 p 由该点处的反射向量确定。源自该贴图的值 $g \circ t$ 随后用于调制颜色强度函数。另外，反射贴图基于相机的位置。

相应地，存在多种技术可执行纹理空间中的、基于反射向量的贴图操作。Blinn 和 Newell 方法考查反射向量与包含当前环境的球体间的交点，并使用球体的参数方程来定义投影 p。此外，也可使用立方体执行该操作。

反射贴图可视为光线跟踪方法的近似操作，其优点主要体现在，可将反射贴图整合至纹理空间内，进而模拟漫反射光照。

例程 shiny_surface 实现了可反射当前环境的材质。

第 17 章 贴图技术

```
298b <shiny surface 298b>≡
  Color shiny_surface(RContext *rc)
  {
    Color ce = environment_map(rc->m->tinfo, reflect_dir(rc->v, rc->n));
    return c_add(c_scale(rc->m->ka, ambient(rc)),
            c_scale(rc->m->ks, c_add(ce, specular(rc))));
  }
Defines:
  shiny_surface, never used.
Uses environment_map 299a.
```

例程 environment_map 利用极坐标实现了反射贴图。

```
299a <environment map 299a>≡
  Color environment_map(void *info, Vector3 r)
  {
    TextureSrc *src = info;

    Vector3 t = sph_coord(r);
    t,x = (t,x / PITIMES2) + 0,5; t,y = (t,y / PI) + 0,5;

    return (*src->texfunc)(src->texdata, t);
  }
Defines:
  environment_map, used in chunk 298b.
Uses sph_coord 299b and TextureSrc 291a.
```

例程 sph_coord 执行直角坐标至极坐标间的转换。

```
299b <sph coord 299b>≡
  Vector3 sph_coord(Vector3 r)
  {
    Real len = v3_norm(r);
    Real theta = atan2(r,y/len, r,x/len);
    Real phi = asin(r,z/len);

    return v3_make(theta, phi, len);
  }
Defines:
  sph_coord, used in chunk 299a.
```

图 17.4 显示了一个环境（或反射）贴图示例。

图 17.4 环境（或反射）贴图

17.6 光源贴图

本节将通过一个光源示例模拟一部幻灯机。

结构 TslideInfo 存储了贴图信息。

```
299c <tslide info 299c>≡
  typedef struct TslideInfo {
    TextureSrc *src;
    Vector3 u, v;
  } TslideInfo;
Defines:
  TslideInfo, used in chunks 300 and 301.
Uses TextureSrc 291a.
```

例程 slide_projector 实现了光源。

```
300 <slide projector 300>≡
  int slide_projector(Light *l, Cone recv, RContext *rc)
  {
    Vector3 c, v, t, m;
    TslideInfo *ti = l->tinfo;

    if (point_coupling(recv, cone_make(l->loc, l->dir, l->cutoff)) == FALSE)
      return FALSE;

    v = v3_sub(rc->p, l->loc);
```

```
    m = v3_make(v3_dot(v, ti->u), v3_dot(v, ti->v), v3_dot(v, l->dir));
    t = v3_make(m,x / m,z * l->distr, m,y / m,z * l->distr, m,z);
    t,x = t,x * 0,5 + 0,5; t,y = t,y * 0,5 + 0,5;
    c = (*ti->src->texfunc)(ti->src->texdata, t);

    l->outdir = v3_scale(-1, v3_unit(v));
    l->outcol = c_scale(l->intensity, c);
    return TRUE;
  }
Defines:
  slide_projector, used in chunk 301.
Uses TslideInfo 299c.
```

例程 slideproj_parse 定义了场景描述语言中的光源。

```
301 <parse slide projector 301>≡
  Val slideproj_parse(int pass, Pval *pl)
  { Val v;
    if (pass == T_EXEC) {
      Light *l = NEWSTRUCT(Light); TslideInfo *ti =
      NEWSTRUCT(TslideInfo);
      l->type = LIGHT_DISTANT; l->color = C_WHITE; l->ambient = 0,1;
      l->intensity = pvl_get_num(pl, "intensity", 1);
      l->cutoff = (DTOR * pvl_get_num(pl, "fov", 90))/2,0;
      l->distr = 1/tan(l->cutoff);
      l->loc = pvl_get_v3(pl, "from", v3_make(0,0,-1));
      l->dir = pvl_get_v3(pl, "at", v3_make(0,0,0));
      l->dir = v3_unit(v3_sub(l->dir, l->loc));
      l->transport = slide_projector; l->tinfo = ti;
      ti->src = parse_texsource(pl);
      ti->v = v3_unit(v3_cross(l->dir, v3_cross(l->dir,
      v3_make(0,1,0))));
      ti->u = v3_cross(l->dir, ti->v);
      v,u,v = l; v,type = LIGHT;
    } return v;
  }
Defines:
  slideproj_parse, never used.
Uses parse_texsource 291b, slide_projector 300, and TslideInfo 299c.
```

图 17.5 显示了基于投影的光源贴图示例。

图 17.5　基于投影的光源贴图

17.7　补充材料

MAP 库中的 **API** 包含了以下例程：

```
Val textured_parse(int pass, Pval *pl);
Val imagemap_parse(int pass, Pval *pl);
TextureSrc *parse_texsource(Pval *pl);

Color textured_plastic(RContext *rc);

Color texture_default();
Color image_texture(Image *i, Vector3 t);

Color texture_map(void *info, Vector3 t);
Color bump_map(void *info, Vector3 t, Vector3 n, Vector3 ds, Vector3 dt);
Color environment_map(void *info, Vector3 r);

Color rough_surface(RContext *rc);
Val rough_parse(int pass, Pval *pl);

Color shiny_surface(RContext *rc);
Val shiny_parse(int pass, Pval *pl);

Real mod(Real a, Real b);
Color chequer(Real freq, Color a, Color b, Vector3 t);

int slide_projector(Light *l, Cone recv, RContext *rc);
Val slideproj_parse(int pass, Pval *pl);
```

第 18 章 着色机制

本章将讨论着色函数的实现过程。

18.1 着色函数采样和重构

着色函数定义于图像平面中,并通过场景可见表面上的光照函数投影得到。同时,着色函数计算可视为采样和重构处理过程。其中,采样过程包括估算可见表面点(对应于图像中的像素)处的光照函数。重构则是指针对图像的其他像素的某些像素处的函数已知值的插值计算。

首先,图像将被划分为多个区域,进而分离处理过程:针对区域边界处的像素,可执行着色函数的采样;随后,对于每个区域内的像素,执行函数的插值计算。另外,划分类型取决于场景对象的几何描述以及光栅化方法。需要注意的是,插值计算将通过光栅化例程完成。对于采用参数方式以及多边形网格描述的对象,图像将被分解为多边形区域。

18.2 采样方法

着色函数的采样可通过点采样或区域采样予以实现。本节仅讨论点采样,区域采样常用于抗锯齿方法中。

例程 point_shade 执行点 p 处的、着色函数的采样。

```
304a <pointshade 304a>≡
  Color point_shade(Vector3 p, Vector3 n, Vector3 v, RContext *rc,
                    Material *m)
  {
    return (*m->luminance)(rc_set(rc, v3_unit(v3_sub(v, p))), p, n, m));
  }
Defines:
  point_shade, used in chunks 304b and 306c.
Uses Material 262a, rc_set, and RContext 259b.
```

18.3 基本的重构方法

用于着色的重构方法包括以下方面。
- Bouknight 着色：分段式常量。
- Gouraud 着色：颜色的线性插值计算。
- Phong 着色：法线的线性插值计算。

18.3.1 Bouknight 着色

例程 flat_shade 对多边形形心处的着色函数进行采样。该计算结果用作图像上多边形区域的常量颜色值。

```
304b <flatshade 304b>≡
  Color flat_shade(Poly *p, Vector3 v, RContext *rc, Material *m)
  {
    Vector3 c = poly_centr(p);
    Vector3 n = poly_normal(p);
    return point_shade(c, n, v, rc, m);
  }
Defines:
  flat_shade, never used.
Uses Material 262a, point_shade 304a, and RContext 259b.
```

18.3.2 Gouraud 方法

结构 GouraudData 存储了 Gouraud 着色方法所用的数据。

```
304c <gourauddata 304c>≡
  typedef struct GouraudData {
    Image *img;
    Poly *cols;
  } GouraudData;
Defines:
  GouraudData, used in chunk 305.
```

例程 gouraud_set 初始化 GouraudData 结构。

第 18 章 着 色 机 制

```
305a <gouraudset 305a>≡
  void *gouraud_set(GouraudData *g, Poly *c, Image *i)
  {
    g->img = i; g->cols = c;
    return (void *)g;
  }
Defines:
  gouraud_set, never used.
Uses GouraudData 304c.
```

例程 gouraud_shade 评估每个多边形顶点处的光照函数。另外，每个顶点处的颜色值将存储于多边形 c 中。该例程应在光栅化之前使用。经该例程计算后的数据，即多边形 c，可通过例程 gouraud_set 进行存储。

```
305b <gouraudshade 305b>≡
  void gouraud_shade(Poly *c, Poly *p, Poly *n, Vector3 v,
                     RContext *rc, Material *m)
  {
    int i;
    for (i = 0; i < p->n; i++)
      c->v[i] =(*m->luminance)(rc_set(rc,v3_unit(v3_sub(v,p->v[i] )),
                        p->v[i],n->v[i],m));
  }
Defines:
  gouraud_shade, never used.
Uses Material 262a, rc_set, and RContext 259b.
```

自图像上多边形区域顶点处的颜色开始，例程 gouraud_paint 重构着色函数。对此，该例程使用了双线性插值计算。需要注意的是，在多边形区域的各像素处，gouraud_paint 将被光栅化例程所调用。

```
305c <gouraudpaint 305c>≡
  void gouraud_paint(Vector3 p, int lv, Real lt, int rv, Real rt, Real st,
                     void *data)
  {
    GouraudData *d = data;
    Vector3 c = seg_bilerp(d->cols, st, lv, lt, rv, rt);
    img_puti(d->img, p.x, p.y, col_dpymap(c));
  }
Defines:
  gouraud_paint, never used.
Uses col_dpymap 312 and GouraudData 304c.
```

18.3.3 Phong 方法

结构 PhongData 存储了 Phong 着色方法所用的数据。

```
306a <phong data 306a>≡
  typedef struct PhongData {
    Poly *pnts;
    Poly *norms;
    Vector3 v;
    RContext *rc;
  } PhongData;
Defines:
  PhongData, used in chunk 306.
Uses RContext 259b.
```

例程 phong_set 初始化结构 PhongData。

```
306b <phongset 306b>≡
  void *phong_set(PhongData *d, Poly *p, Poly *n, Vector3 v,
                  RContext *rc, Material *m)
  {
    d->pnts = p;
    d->norms = n;
    d->v = v;
    d->rc = rc;
    d->rc->m = m;
    return (void *)d;
  }
Defines:
  phong_set, never used.
Uses Material 262a, PhongData 306a, and RContext 259b.
```

例程 phong_shadepaint 计算多边形顶点处的法线并评估光照函数。

```
306c <phongshadepaint 306c>≡
  void phong_shadepaint(Vector3 p, int n, int lv, Real lt,
                        int rv, Real rt, Real st, void * data)
  {
    PhongData *d = data;
    Vector3 pv = seg_bilerp(d->pnts, n, st, lv, lt, rv, rt);
    Vector3 pn = v3_unit(seg_bilerp(d->norms, n, st, lv, lt, rv, rt));
    Color c = point_shade(pv, pn, d->v, d->rc, d->rc->m);
    img_putc(d->rc->img, p.x, p.y, col_dpymap(c));
```

```
}
Defines:
  phong_shadepaint, never used.
Uses col_dpymap 312, PhongData 306a, and point_shade 304a.
```

18.4 纹理属性的重构

上述内容利用 Gouraud 和 Phong 方法重构了着色函数。此类方法采用了线性插值在每个像素处重构与场景多边形顶点关联的属性值。

线性投影仅对基于正交投影的相机转换生成正确的重构结果。对于投影相机转换（透视投影），线性插值将产生错误的结果。尽管如此，线性插值仍被广泛地应用，其原因在于，从感知角度来看，误差并不十分明显，属性函数仅包含少量变化，如颜色和法线。

然而，对于包含高频的属性函数来说，如纹理，误差则表现得十分明显，且有必要使用投影重构方法，稍后将对此加以讨论。

18.4.1 插值和投影转换

这里首先回顾一下与投影转换相关的概念。当把场景表面的一点 $p \in \mathbb{R}^3$ 映射至图像平面的投影 $s \in \mathbb{R}^2$ 中时，可使用下列投影转换：

$$s = Mp$$

对应的矩阵形式如下所示：

$$\begin{pmatrix} uw \\ vw \\ \dots \\ w \end{pmatrix} = \begin{pmatrix} e_1 & e_2 & \dots & e_n \\ f_1 & f_2 & \dots & f_n \\ \dots \\ g_1 & g_2 & \dots & g_n \end{pmatrix} \cdot \begin{pmatrix} x \\ y \\ \dots \\ 1 \end{pmatrix}$$

其中，M 表示为投影转换矩阵，p 和 s 则采用齐次坐标表示。

点 $p = (x, y, \dots, 1)$ 属于内嵌于投影空间 $\mathbb{RP}^3$ 中的仿射超平面。当把点 $s = (uw, vw, \dots, w)$ 转换至标准化仿射点 $\bar{s} = (u, v, \dots, 1)$ 时，需要执行 w 齐次除法。随后可得到：

$$u = \frac{e_1 x + e_2 y + \dots + e_n}{g_1 x + g_2 y + \dots + g_n}$$

而对于其他坐标 $(u, v, \dots)$，也可得到类似的结果。

屏幕平面与世界空间的逆转换如下所示：

$$p = M^{-1} s$$

或者

$$\begin{pmatrix} x \\ y \\ \dots \\ 1 \end{pmatrix} = \begin{pmatrix} a_1 & a_2 & \dots & A \\ b_1 & b_2 & \dots & B \\ \dots & & & \\ c_1 & c_2 & \dots & C \end{pmatrix} \begin{pmatrix} uw \\ vw \\ \dots \\ w \end{pmatrix}$$

其中，p 的坐标 x 可计算为：

$$x = \frac{(a_1 u + b_1 v + \dots + c_1)w}{(A + B + \dots + C)w} = \frac{a_1 u + b_1 v + \dots + c_1}{A + B + \dots + C}$$

对于其他坐标也可得到类似的结果。

需要注意的是，世界空间中的属性值可通过重构函数表示，该函数采用屏幕坐标 $s = (u, v, \dots, w)$ 进行参数化，例如，x 可通过下列公式得到：

$$x(s) = \frac{a_1 u + \dots + c_1}{A + \dots + C}$$

换而言之，这说明了针对投影相机转换，正确的重构函数应使用线性有理插值。

图 18.1 显示了线性插值和有理插值之间的区别。

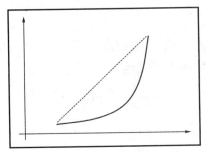

图 18.1　线性（虚线）和有理（实线）插值

可以看到，Gouraud 和 Phong 方法采用了属性的线性插值，并包含了下列结构。

（1）计算 $x(s_0)$ 和 $x(s_k)$。

（2）线性插值：$x(s) - (1 - t)x(s_0) + tx(s_k)$。

我们已经知道，该模式将生成错误的结果，如果 $x(s)$ 包含高频内容，即可感知到这一错误结果。

一种处理此类问题的方法是，插值参数 s，并针对每个像素利用逆映射计算属性值，如下所示。

（1）线性插值：$s = s_0, \dots s_k$。

（2）评估 $x(s) = M^{-1}s$。

第18章 着色机制

该方法尽管正确,但却相对低效——每个像素均会执行矩阵和向量间的乘法运算。一种高效的解决方案是将有理线性插值用作重构函数,如下所示:

(1) 计算 s_0 和 s_k 中的 $x_w = \dfrac{x}{w}$, $d_w = \dfrac{1}{w}$。

(2) 对 $x_w(s)$ 和 $d_w(s)$ 执行线性插值,随后得到 $x(s) = \dfrac{x_w(s)}{d_w(s)}$。

注意,w 值已通过直接投影转换 $s = (u\ldots w)^T = Mp$ 计算完毕。

为了证明上述结果的正确性,考查下列插值结果:

$$\frac{x(s)}{w(s)} = a_1 u + \ldots + c_1, \quad \frac{1}{w(s)} = A + \ldots + C$$

对应结果为线性表达式。

针对每个像素 s 执行除法运算,可得到:

$$x(s) = \frac{x(s)/w(s)}{1/w(s)} = \frac{a_1 u + \ldots + c_1}{A + \ldots + C}$$

这也是我们期望的结果。

18.4.2 纹理的有理线性插值

当采用有理线性插值进行重构时,首先需要分离与每个场景多边形顶点关联的齐次坐标 w。随后,执行插值计算的全部属性均除以 w。最后,这一类属性传递至光栅化例程中,执行插值计算以及每个像素最终的除法运算。

例程 poly_wz_hpoly 针对每个多边形顶点计算坐标 w,并将值 $1/w$ 置于某个属性多边形中。

```
309 <poly wz hpoly 309>≡
 Poly *poly_wz_hpoly(Poly *q, Hpoly *s)
 {
   Poly *w; int i;
   for (i = 0; i < s->n; i++) {
     q->v[i] = v3_make(s->v[i].x, s->v[i].y, s->v[i].z);
     if (w = q->v->next) {
       Real rw = 1./s->v[i].w;
       w->v[i] = v3_make(rw, rw, rw);
     }
   }
   return q;
 }
```

例程 texture_wscale 将全部纹理属性乘以 1/w。

```
310a <texture wscale 310a>≡
  Poly *texture_wscale(Poly *w, Poly *p)
  {
    Poly *l; int i;
    for (i = 0; i < w->n; i++)
      for (l = p; l != NULL; l = l->next) {
        l->v[i] = v3_scale(w->v[i].z, l->v[i] );
      }
    return w;
  }
Defines:
  texture_wscale, never used.
```

数据结构 TextureData 包含了插值后的属性 vpar。

```
310b <texture data 310b>≡
  typedef struct TextureData {
    Image *img;
    Vector3 eye;
    Poly *vpar;
    RContext *rc;
  } TextureData;
Defines:
  TextureData, used in chunks 310c and 311.
Uses RContext 259b.
```

例程 texture_set 初始化 TextureData 结构，并存储属性和其他数据。

```
310c <texture set 310c>≡
  void *texture_set(TextureData *d, Poly *param, Vector3 eye,
                    RContext *rc, Light *l, Material *m, Image *i)
  { int n;
    if ((n = plist_lenght(param)) < 1|| n > TEX_MAXPAR)
      fprintf(stderr, "Texture: not enough parameters\n");
    d->img = i;
    d->vpar = param;
    d->eye = eye;
    d->rc = rc;
    d->rc->l = l;
    d->rc->m = m;
```

```
    return (void *)d;
  }
Defines:
 texture_set, never used.
Uses Light 258a, Material 262a, RContext 259b, and TextureData 310b.
```

例程 texture_shadepaint 执行有理线性插值计算并调用光照函数,这将使用到纹理坐标。

```
311 <texture shadepaint 311>≡
  void texture_shadepaint(Vector3 s, int lv, Real lt, int rv, Real rt, Real
                          st,void *data)
  {
    Vector3 a[TEX_MAXPAR+1] ;
    Poly *l; Color c; int i;
    TextureData *d = data;

    a[TEX_W] = seg_bilerp(d->vpar, st, lv, lt, rv, rt);
    for (l = d->vpar->next, i = 1; l != NULL; l = l->next, i++)
      a[i] = v3_scale(1/a[TEX_W].z, seg_bilerp(l,st,lv,lt,rv,rt));
    a[TEX_N] = v3_unit(a[TEX_N] );
    a[TEX_E] = v3_unit(v3_sub(d->eye, a[TEX_P] ));
    c = (*d->rc->m->luminance)(rc_tset(d->rc, a[TEX_E], a[TEX_P], a[TEX_N],
                               a[TEX_T], a[TEX_U],a[TEX_V] ));
    img_puti(d->img, s.x, s.y, col_dpymap(c));
  }
Defines:
 texture_shadepaint, never used.
Uses col_dpymap 312, rc_tset, and TextureData 310b.
```

上述实现定义了下列属性:

```
#define TEX_W 0              // 1/w
#define TEX_T 1              // texture coordinates
#define TEX_P 2              // position on surface
#define TEX_N 3              // normal to P
#define TEX_U 4              // partial derivative df/du
#define TEX_V 5              // partial derivative df/dv
#define TEX_E 6              // vector towards observer
```

18.5 图 像 化

一旦着色函数计算完毕,可将计算后的数值映射至图像的颜色空间中。计算光照所用的颜色表达方式通过结构 Color 指定，该结构在 Vector3 结构的基础上加以定义，同时整合了向量操作。运算 c_mult 则单独予以实现。

```
#define Color Vector3

#define c_add(a, b) v3_add(a,b)
#define c_sub(a, b) v3_sub(a,b)
#define c_scale(a, b) v3_scale(a, b)
#define c_sqrnorm(a) v3_sqrnorm(a)

Color c_mult(Color a, Color b);
```

针对基本颜色（白色和黑色）的构建，此处定义了相应的宏，如下所示：

```
#define C_WHITE v3_make(1,1,1)
#define C_BLACK v3_make(0,0,0)
```

例程 col_dpymap 将标准化颜色值（颜色分量位于 0～1 范围）转换为某种颜色，对应的分量位于 0～255 范围内。

```
312 <colordpy map 312>≡
  Pixel col_dpymap(Color c)
  {
    double gain = 255.0;
  #ifdef RGB_IMAGE
    return rgb_to_index(c_scale(gain, c));
  #else
    return rgb_to_y(c_scale(gain, c));
  #endif
  }
Defines:
  col_dpymap, used in chunks 305c, 306c, and 311.
```

18.6　补　充　材　料

SHADE 库的 API 包含了以下例程：

```
Color point_shade(Vector3 p,Vector3 n,Vector3 v, RContext *rc, Light *l,
              Material *m);
Color flat_shade(Poly *p, Vector3 v, RContext *rc, Light *l, Material *m);
Color gouraud_paint(Vector3 p, int lv, Real lt, int rv, Real rt, Real st,
              void *data);
```

```
void gouraud_shade(Poly *p, Poly *n, Poly *c, Vector3 v, RContext *rc,
            Light *l, Material *m);

Color phong_shade(Vector3 p, int lv, Real lt, int rv, Real rt, Real st,
            void *data);

Poly *texture_wscale(Poly *w, Poly *p);
Poly *poly_wz_hpoly(Poly *q, Poly *w, Hpoly *s);

void *texture_set(TextureData *d, Poly *param, Vector3 eye,
            RContext *rc, Light *l, Material *m, Image *i);
void texture_shadepaint(Vector3 p, int lv, Real lt, int rv, Real rt, Real st,
            void *data);

Pixel col_dpymap(Color c);
```

第 19 章　三维图形系统

本章介绍如何将本书中的计算机图形学算法整合至 3D 图形系统中。该系统由两个子系统构成，即建模系统和渲染系统。下面将展示 3 种三维系统架构，并通过一种自然的方式整合这两个子系统中的各种技术，具体如下。

- 系统 A：生成模型+基于 Z-缓冲区的渲染机制。
- 系统 B：CSG 建模+基于光线跟踪的渲染机制。
- 系统 C：基于图元的建模机制+基于 Painter 算法的渲染机制。

接下来将讨论上述 3 种系统的实现。在基础版本中，系统将通过命令行方式并以非交互模式运行。相应地，交互式系统将作为本章结束时的项目实践内容。

19.1　系统 A

系统 A 基于生成模型和 Z-缓冲区渲染机制。

19.1.1　生成模型

建模子系统 rotsurf 生成的旋转曲面近似于多边形网格。其中采用了参数化的几何描述、边界分解表达模式，以及生成式建模技术。模型器接收多边形曲线作为输入内容，并输出三角形网格。

主程序将读取输入数据并调用例程 rotsurf 和 trilistwrite 构建旋转曲面，并写入多边形网格中。

```
313 <rotsurf 313>≡
 #define MAXPTS 2048
 Vector3 g[MAXPTS] ;
 main(int argc, char **argv)
 {
   int nu, nv = NVPTS;
   Poly *tl;
   if (argc == 2)
     nv = atoi(argv[1] );
   nu = read_curve();
```

```
    tl = rotsurf(nu, g, nv);
    trilist_write(tl, stdout);
    exit(0);
  }
Defines:
  g, used in chunks 314, 321b, and 322b.
  MAXPTS, used in chunk 314.
Uses main 47 187b 315 321b and read_curve 314.
```

例程 read_curve 读取多边形曲线的顶点,并将其存储至全局数组 g 中。

```
314 <readcurve 314>≡
  int read_curve(void)
  {
    int k = 0;
    while (scanf("%lf %lf %lf\n",&(g[k].x),&(g[k].y),&(g[k].z)) != EOF)
      if (k++ > MAXPTS)
        break;
    return k;
  }
Defines:
  read_curve, used in chunk 313.
Uses g 313 and MAXPTS 313.
```

例如,命令 rotsurf 12 < ln.pts > cyl.scn 将构建包含 12 条边的圆柱面的多边形网格。输入文件 ln.pts 将在平面 $z = 0$ 中指定一行。

```
1 1 0
1 -1 0
```

输出文件 cyl.out 包含了表面多边形列表(为了节省篇幅,此处仅显示了部分内容)。

```
trilist {
{{0.186494, -1, 0.0722484}, {0.2, -1, 0}, {1, -1, 0}},
{{0.932472, -1, 0.361242}, {0.186494, -1, 0.0722484}, {1, -1, 0}},
{{0.932472, -1, 0.361242}, {1, -1, 0}, {1, 1, 0}},
{{0.932472, 1, 0.361242}, {0.932472, -1, 0.361242}, {1, 1, 0}},
{{0.932472, 1, 0.361242}, {1, 1, 0}, {0.2, 1, 0}},
{{0.186494, 1, 0.0722484}, {0.932472, 1, 0.361242}, {0.2, 1, 0}},
{{0.147802, -1, 0.134739}, {0.186494, -1, 0.07224}, {0.93247, -1, 0.361}},
{{0.739009, -1, 0.673696}, {0.147802, -1, 0.13473}, {0.93247, -1, 0.361}},
....
```

{{0.2, 1, 0}, {1, 1, 0}, {0.186494, 1, -0.0722483}}}
}

19.1.2 基于 Z-缓冲区的渲染机制

渲染器 zbuff 采用了渐增式光栅化，Z-缓冲区负责执行可见性计算。另外，光照则使用了基于面元着色的漫反射 Lambert 模型。

全局数据结构 s、rc、mclip 和 mdpy 分别存储了场景对象列表、渲染上下文和两个相机转换，如下所示：

```
static Scene *s;
static RContext *rc;
static Matrix4 mclip, mdpy;
```

主程序将读取场景数据、初始化全局结构并处理对象列表。针对每个朝向相机的多边形，将计算多边形形心处的光照函数。随后，多边形将被转换，并通过 Z-缓冲区执行剪裁和光栅化操作。在处理结束时，所生成的图像将被写入输出文件中。

```
315 <zbuff 315>≡
  int main(int argc, char **argv)
  {
    Object *o; Poly *p; Color c;

    init_sdl();
    s = scene_read();
    init_render();
    for (o = s->objs; o != NULL; o = o->next) {
      for (p = o->u.pols; p != NULL; p = p->next) {
        if (is_backfacing(p, v3_sub(poly_centr(p), s->view->center)))
          continue;
        c = flat_shade(p, s->view->center, rc, o->mat);
        if(poly_clip(VIEW_ZMIN(s->view),poly_transform(p,mclip),0))
          scan_poly(poly_homoxform(p, mdpy), pix_paint, &c);
      }
    }
    img_write(s->img, "stdout", 0);
    exit(0);
  }
Defines:
  main, used in chunks 313, 317, and 318c.
Uses init_render 316c 319b 322b, init_sdl 316b 319c 322a, and pix_paint 316a.
```

例程 pix_paint 利用 Z-缓冲区绘制图像。

316a <pix paint 316a>≡
```
 void pix_paint(Vector3 v,int n,int lv,Real lt,int rv,Real rt,Real st,
                void *c)
 {
   if (zbuf_store(v))
     img_putc(s->img, v.x, v.y, col_dpymap(*((Color *)(c))));
 }
```
Defines:
  pix_paint, used in chunk 315.

例程 init_sdl 负责注册场景描述语言的运算符。

316b <sdlzbuff 316b>≡
```
 void init_sdl(void)
 {
   lang_defun("scene", scene_parse);
   lang_defun("view", view_parse);
   lang_defun("dist_light", distlight_parse);
   lang_defun("plastic", plastic_parse);
   lang_defun("polyobj", obj_parse);
   lang_defun("trilist", trilist_parse);
 }
```
Defines:
  init_sdl, used in chunks 315, 318c, and 321b.

例程 init_render 初始化转换矩阵、Z-缓冲区和渲染上下文。

316c <initzbuff 316c>≡
```
 void init_render(void)
 {
   mclip = m4_m4prod(s->view->C, s->view->V);
   mdpy = m4_m4prod(s->view->S, s->view->P);
   zbuf_init(s->img->w, s->img->h);
   rc_sset(rc = NEWSTRUCT(RContext), s->view, s->lights, s->img);
 }
```
Defines:
  init_render, used in chunks 315, 318c, and 321b.

命令 zbuff < cyl.scn > cyl.ras 根据场景描述文件生成一幅图像。注意，这里省略了部分多边形列表。

316d <cyl scn 316d>≡
```
 scene {
```

```
        camera = view { from = {0, 0, -2.5}, up = {0, 1, 0}},
        light = dist_light {direction = {0, 0, -1} },
        object = polyobj { shape = trilist {
                    {{0.186494, -1, 0.0722484}, {0.2, -1, 0}, {1, -1, 0}}
                    ...
                    {{0.2, 1, 0}, {1, 1, 0}, {0.186494, 1, -0.0722483}}}
        }
};
```

图 19.1 显示了表面的渲染结果,即图像 cyl.ras。

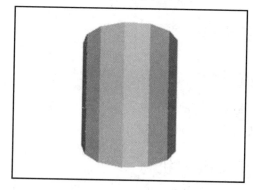

图 19.1　多边形模型的渲染结果

19.2　系统 B

系统 B 基于 CSG 建模机制以及基于光线跟踪的可视化操作。

19.2.1　CSG 建模机制

建模器 csg 将解释构建式实体模型。其中,图元表示为球体,表达方式为 CSG 表达式,建模技术则基于语言实现。

主程序读取一个 CSG 表达式,并将对应的构建结果写入场景描述语言中。

```
317 <csg 317>≡
  main(int argc, char **argv)
  {
    CsgNode *t;
    if((t = csg_parse()) == NULL)
      exit(-1);
```

```
      else
        csg_write(t, stdout);
      exit(0);
    }
Uses main 47 187b 315 321b.
```

例如，命令 csg < s.csg > s.scn 将 s.csg 中的 CSG 表达式转换至 s.scn。文件 s.csg 中的对象由两个球体间的差值构成。

```
318a <spheres csg 318a>≡
  (s{ 0 0 0 1 } \ s{ 1 1 -1 1 })
```

文件 s.scn 中包含了场景描述语言中的相关命令。

```
318b <sphere scn 318b>≡
  csgobj = csg_diff {
                     csg_prim{ sphere { center = {0, 0, 0}}},
                     csg_prim{ sphere { center = {1, 1, -1}}} }
```

当前程序可作为开发完整建模器的基础内容，后者一般会包含编辑 CSG 对象的操作。

19.2.2 基于光线跟踪的渲染机制

渲染器 rt 采用了外在型光栅化操作，并通过光线跟踪机制执行可见性计算。相应地，光照基于 Phong 镜面模型，并利用点采样着色。

主程序读取场景数据、初始化全局结构并生成图像。针对图像上的每一个像素，场景中将投射一条光线，同时计算该光线与 CSG 对象间的交点。针对最近表面的交点，将进一步确定光照函数值。最后，图像写入输出文件中。

```
318c <rt 318c>≡
  main(int argc, char **argv)
  {
    Color c; int u, v;
    Ray r; Inode *l;
    init_sdl();
    s = scene_read();
    init_render();
    for (v = s->view->sc.ll.y; v < s->view->sc.ur.y; v += 1) {
      for (u = s->view->sc.ll.x; u < s->view->sc.ur.x; u += 1) {
        r = ray_unit(ray_transform(ray_view(u, v), mclip));
        if ((l = ray_intersect(s->objs, r)) != NULL)
          c = point_shade(ray_point(r, l->t), l->n, s->view->center,
                          rc,l->m);
```

```
        else
          c = bgcolor;
        inode_free(l);
        img_putc(s->img, u, v, col_dpymap(c));
      }
    }
    img_write(s->img,"stdout",0);
    exit(0);
  }
Uses init_render 316c 319b 322b, init_sdl 316b 319c 322a, main 47 187b 315
321b, and ray_view 319a.
```

例程 ray_view 构建一条光线，该光线离开虚拟相机，并穿越图像平面上坐标为(u, v) 的像素。

```
319a <ray view 319a>≡
  Ray ray_view(int u, int v)
  {
    Vector4 w = v4_m4mult(v4_make(u, v, s->view->sc.ur.z, 1), mdpy);
    return ray_make(v3_v4conv(v4_m4mult(v4_make(0, 0, 1, 0), mdpy)),
                    v3_make(w.x, w.y, w.z));
  }
Defines:
  ray_view, used in chunk 318c.
```

例程 init_render 初始化相机转换操作。

```
319b <init rt 319b>≡
  void init_render(void)
  {
    mclip = m4_m4prod(s->view->Vinv, s->view->Cinv);
    mdpy = m4_m4prod(s->view->Pinv, s->view->Sinv);
    rc_sset(rc = NEWSTRUCT(RContext), s->view, s->lights, s->img);
  }
Defines:
  init_render, used in chunks 315, 318c, and 321b.
```

例程 init_sdl 负责注册场景描述语言中的运算符。

```
319c <sdl rt 319c>≡
  void init_sdl(void)
  {
    lang_defun("scene", scene_parse);
    lang_defun("view", view_parse);
```

```
    lang_defun("dist_light", distlight_parse);
    lang_defun("plastic", plastic_parse);
    lang_defun("csgobj", obj_parse);
    lang_defun("csg_union", csg_union_parse);
    lang_defun("csg_inter", csg_inter_parse);
    lang_defun("csg_diff", csg_diff_parse);
    lang_defun("csg_prim", csg_prim_parse);
    lang_defun("sphere", sphere_parse);
}
Defines:
    init_sdl, used in chunks 315, 318c, and 321b.
```

作为渲染器应用示例，命令 rt < s.scn > s.ras 生成了文件 s.scn 中描述的、CSG 场景的一幅图像。

```
320 <csg2 scn 320>≡
  scene{
    camera = view {
        from = {0, 0, -4}, at = {0, 0, 0}, up = {0,1,0}, fov = 60},
    light = dist_light {direction = {0, 1, -1} },
    object = csgobj{
        material = plastic { ka = .2, kd = 0.8, ks = 0.0 },
        shape = csg_diff {
                csg_prim{ sphere { center = {0, 0, 0}}},
                csg_prim{ sphere { center = {1, 1, -1}}}
            }
        }
    }
}
```

图 19.2 显示了场景的渲染结果，即图像 s.ras。

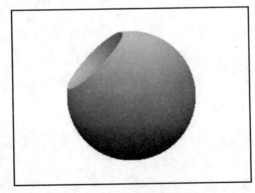

图 19.2 CSG 场景渲染

19.3 系统 C

系统 C 基于图元建模，并利用 Painter 方法进行渲染。

19.3.1 基于图元层次结构的建模

建模器使用了两种形式表述的几何图元，即参数式和隐式，并通过仿射转换以及层次结构方式进行分组。稍后将查看场景描述语言中图元的层次结构示例。

```
321a <hierscn 321a>≡
 hier {
        transform { translate = { .5, .5, 0}},
        group {
                transform { zrotate = .4 },
                obj = sphere{ },
                transform { translate = {.2, 0, 1}},
                group {
                        transform{ scale = {2, 0.4, 1}},
                        obj = sphere{ radius = .1 } }
 };
```

19.3.2 基于 Painter 方法的渲染机制

渲染器 zsort 采用 Painter 方法确定可见表面；光照则利用 Gouraud 插值的漫反射模型。全局数据结构 z 和 g 分别存储场景多边形的排序列表，以及当前多边形 Gouraud 插值的数据。

```
static Scene *s;
static Object *o;
static Matrix4 mclip, mdpy;
static RContext *rc;
static List *z = NULL;
static GouraudData *g;
```

主程序读取场景数据、初始化相关结构，并执行多边形剪裁和移除操作。相机视域内的多边形在 Z 缓冲区中排序并被光栅化。

321b <zsort 321b>≡
```
  int main(int argc, char **argv)
  {
    Poly *l, *p, *c = poly_alloc(3);
    Item *i;
    init_sdl();
    s = scene_read();
    init_render();
    for (o = s->objs; o != NULL; o = o->next) {
      for (l = prim_uv_decomp(o->u.prim, 1.); l != NULL; l = l->next) {
        p = poly_transform(prim_polys(o->u.prim, l), mclip);
        if (! is_backfacing(p, v3_unit(v3_scale(-1, poly_centr(p)))))
          hither_clip(VIEW_ZMIN(s->view), p, z_store, plist_free);
      }
    }
    z = z_sort(z);
    for (i = z->head; i != NULL; i = i->next) {
      gouraud_shade(c, P(i), N(i), s->view->center, rc, M(i));
      p = poly_homoxform(S(i),mdpy);
      scan_poly(p, gouraud_paint, gouraud_set(g,s->img));
    }
    img_write(s->img, "stdout", 0);
    exit(0);
  }
```
Defines:
  main, used in chunks 313, 317, and 318c.
Uses g 313, init_render 316c 319b 322b, init_sdl 316b 319c 322a, prim_polys 323a, and z_store 322c.

例程 init_sdl 负责注册场景描述语言中的运算符。

322a <sdl zsort 322a>≡
```
  void init_sdl(void)
  {
    lang_defun("scene", scene_parse);
    lang_defun("view", view_parse);
    lang_defun("dist_light", distlight_parse);
    lang_defun("plastic", plastic_parse);
    lang_defun("primobj", obj_parse);
    lang_defun("sphere", sphere_parse);
  }
```
Defines:
  init_sdl, used in chunks 315, 318c, and 321b.

第 19 章 三维图形系统

例程 init_render 初始化全局数据结构。

```
322b <initzsort 322b>≡
  void init_render(void)
  {
    mclip = m4_m4prod(s->view->C, s->view->V);
    mdpy = m4_m4prod(s->view->S, s->view->P);
    z = new_list();
    g = NEWSTRUCT(GouraudData);
    rc_sset(rc = NEWSTRUCT(RContext), s->view, s->lights, s->img);
  }
Defines:
  init_render, used in chunks 315, 318c, and 321b.
Uses g 313.
```

例程 z_store 将多边形存储于 z 列表中。

```
322c <zstore 322c>≡
  void z_store(Poly *l)
  {
    Zdatum *d = NEWSTRUCT(Zdatum);
    d->zmax = MIN(l->v[0].z, MIN(l->v[1].z, l->v[2].z));
    d->l = l; d->o = o;
    append_item(z, new_item(d));
  }
Defines:
  z_store, used in chunk 321b.
```

下列宏用于访问多边形属性列表。

```
#define S(I)   (((Zdatum *)(I->d))->l)
#define P(I)   (((Zdatum *)(I->d))->l->next)
#define N(I)   (((Zdatum *)(I->d))->l->next->next)
#define M(I)   (((Zdatum *)(I->d))->o->mat)
#define SL(L)  (l)
#define PL(L)  (l->next)
#define NL(L)  (l->next->next)
```

例程 prim_polys 创建包含位置和图元表面点法线的三角形。

```
323a <prim polys 323a>≡
  Poly *prim_polys(Prim *s, Poly *p)
  {
    int i; Poly *l = plist_alloc(3, p->n);
    for (i = 0; i < p->n; i++) {
```

```
        PL(l)->v[i] = SL(l)->v[i] = prim_point(s, p->v[i].x, p->v[i].y);
        NL(l)->v[i] = prim_normal(s, p->v[i].x, p->v[i].y);
    }
    return 1;
}
Defines:
    prim_polys, used in chunk 321b.
```

例如，命令 zsort < pr.scn > pr.ras 生成文件 prim.scn（参见下列代码）中描述的、包含两个球体的场景图像。

```
323b <prim scn 323b>≡
    scene{
        camera = view {
            from = {0, 0, -2.5}, at = {0, .5, 0}, up = {0,1,0}, fov = 90},
        light = dist_light {direction = {0, 1, -1} },
        object = primobj{
            material = plastic { ka = .2, kd = 0.8, ks = 0.0 },
            shape = sphere { center = {0, 0, 0}}},
        object = primobj{
            material = plastic { ka = .2, kd = 0.8, ks = 0.0 },
            shape = sphere { center = {2, 2, 2}}},
    };
```

图 19.3 显示了场景渲染结果，即图像 prim.ras。

图 19.3　通过 Z-排序渲染球体

19.4　项　　目

本节将建立多个项目，并整合本书中所开发的库。如前所述，建模和渲染成语基于系统 A、B、C 架构。

19.4.1 渲染图像的程序

本节将利用 GP、COLOR、IMAGE 库开发图像渲染程序。该程序将接收以下类型的光栅化格式文件。

- 灰度图。
- 真彩图（RGB）。
- 索引颜色。
- RLE 压缩灰度图。
- RLE 压缩索引颜色。

该程序应可显示图像和所关联的颜色图。

下列窗口系统事件将通过程序进行控制：窗口弹出事件和窗口尺寸变化事件。图 19.4 显示了 Mandrill 和 Lenna 测试图像。

图 19.4　Mandrill 和 Lenna 测试图像

19.4.2 建模系统

本节将开发基本的建模系统之一，并从扩展列表中至少实现一种选择方案。前述内容描述的基本系统包括：

- 生成式建模机制。
- CSG 建模机制。
- 图元建模机制。

在每种系统中，应针对对象属性规范（如名称、颜色、材质等）纳入相关引擎。

系统应采用复杂示例进行测试，对此需要：

（1）利用建模器构建复杂度合理的模型。

（2）解释对象的生成方式。

（3）对难点加以讨论。

生成式建模机制包括：

（1）包含 3 种正交（XYZ）视见渲染。

（2）涵盖交互式功能。用户应可在屏幕上绘制一条曲线，并根据该曲线生成一个旋转曲面。

（3）具有曲线编辑功能，如移动一个顶点，或者添加一个被移除的顶点。此类变化应反映于当前表面上。

（4）除了旋转表面之外，还应包含其他生成式模型类型，如挤压、扭曲、弯曲和锥化。

（5）包含与交互式选项集成的命令语言。

CSG 建模机制包括：

（1）包含 3 种正交（XYZ）视见渲染。

（2）包含转换操作（csg transform 运算符）。用户应对图元进行分组和转换。

（3）包含交互式功能。用户应可选择、移动图元或图元组。

（4）包含 CSG 树形结构的渲染机制，并可对其结构进行操控。用户应可在交互式图形模式中创建、编辑 CSG 结构。

（5）包含计算对象属性的功能（如体积等）。

（6）包含与交互式选项集成的命令语言。

基于图元的建模机制包括：

（1）包含 3 种正交（XYZ）视见和辅助视见渲染。

（2）包含编辑命令，如创建、删除、重命名、取消和选择操作等。

（3）包含下列图元：盒体、圆锥体、圆环体和超二次曲面。图元实例应通过符号化名称标识，并包含与其关联的转换操作。用户应可在对象创建完毕后调整其全部参数。

（4）包含对象组。对象组应通过名称进行标识。另外，分组层次结构应可在辅助窗口中渲染。

关节式对象建模机制包括：

（1）包含图元圆柱体。

（2）包含对关节的支持（如球面枢轴类型）。

（3）包含动画序列。

19.4.3 渲染系统

本节将开发基本的渲染系统之一，并从扩展列表中至少实现一种选择方案。前述内容描述的基本系统包括：
- 基于扫描线的渲染机制。
- 基于光线跟踪的渲染机制。
- 基于 Painter 方法的渲染机制。

系统应可通过复杂示例进行测试，生成图像并描述渲染器的特征。

利用程序运行分析编写技术报告，该项工作涉及以下几点内容：

（1）基于程序执行的设定内容，标识并解释算法的瓶颈。

（2）讨论加速程序运行的可选方案。

基于扫描线的渲染机制包括：

（1）利用预处理辐射度包含全局光照。

（2）针对改进后的辐射度计算包含加速方案，如自适应细分。

（3）包含隐式图元和 CSG 模型。

基于光线跟踪的渲染机制包括：

（1）包含基于递归光线跟踪的全局光照。

（2）包含光线跟踪加速方法。

（3）包含多边形对象。

基于 Painter 方法的渲染机制包括：

（1）包含纹理贴图机制。

（2）包含光照贴图机制。

（3）包含 3D 过程式纹理。

（4）包含用于加速的细节级别（LOD）。

（5）使用反向多边形绘制，以提高效率。

参 考 文 献

[Aho and Ullman 79] A. Aho and J. Ullman. *Principles of Compiler Design*. Addison-Wesley, 1979.

[Atherton et al. 78] P. Atherton, K. Weiler, and D. Greenberg. "Polygon Shadow Generation." *Computer Graphics (SIGGRAPH'78 Proceedings)* 12:3 (1978), 275–281.

[Barrow 02] John D. Barrow. *The Constants of Nature: From Alpha to Omega—The Numbers that Encode the DeepestSecrets of the Universe*. New York: Pantheon Books, 2002.

[Blinn and Newell 76] James F. Blinn and Martin E. Newell. "Texture and Reflection in Computer Generated Images." *Communications of the ACM* 19:10 (1976), 542–547.

[Bouknight and Kelly 70] W. J. Bouknight and K. C. Kelly. "An Algorithm for Producing Half-Tone Computer Graphics Presentations with Shadows and Movable Light Sources." In *Proc. AFIPS JSCC*, 36, 36, pp. 1–10. New York: ACM, 1970.

[Carpenter 84] Loren Carpenter. "The A-buffer, an Antialiased Hidden Surface Method." *Com-puter Graphics (SIGGRAPH'84 Proceedings)* 18:3 (1984), 103–108.

[Clark 82] James H. Clark. "The Geometry Engine: A VLSI Geometry System for Graphics." *Computer Graphics (SIGGRAPH'82 Proceedings)* 16:3 (1982), 127–133.

[Cook and Torrance 81] R. L. Cook and K. E. Torrance. "A Reflectance Model for Computer Graphics." *Computer Graphics (SIGGRAPH'81 Proceedings)* 15:3 (1981), 307–316.

[Cook et al. 87] Robert L. Cook, Loren Carpenter, and Edwin Catmull. "The Reyes Image Ren-dering Architecture." *Computer Graphics (SIGGRAPH'87 Proceedings)*, pp. 95–102.

[Crow 82] F. C. Crow. "A More Flexible Image Generation Environment." *Computer Graphics(SIGGRAPH'82 Proceedings)* 16:3 (1982), 9–18.

[Fuchs et al. 83] H. Fuchs, G. D. Abram, and E. D. Grant. "Near Real-Time Shaded Display of Rigid Objects." *Computer Graphics (SIGGRAPH'83 Proceedings)* 17:3 (1983), 65–72.

[Gomes and Velho 95] Jonas Gomes and Luiz Velho. *Computac̜, ão Gráfica: Imagem*. IMPA-SBM, 1995.

[Gomes and Velho 97] Jonas Gomes and Luiz Velho. *Image Processing for ComputerGraphics*. New York: Springer-Verlarg, 1997.

[Gomes and Velho 98] Jonas Gomes and Luiz Velho. *Computac, ão Gráfica: Volume 1*. IMPA-SBM,1998.

[Gomes et al. 96] J. Gomes, L. Darsa, B. Costa, and L. Velho. "Graphical Objects." *The Visual Computer* 12 (1996), 269–282.

[Gomes et al. 12] Jonas Gomes, Luiz Velho, and Mario Costa Sousa. *Computer Graphics: Theory and Practice*. Boca Raton, FL: CRC Press, 2012.

[Heckbert and Hanrahan 84] Paul S. Heckbert and Pat Hanrahan. "Beam Tracing Polygonal Ob-jects." *Computer Graphics (SIGGRAPH'84 Proceedings)* 18:3 (1984), 119–127.

[MAGI 68] MAGI. "3-D Simulated Graphics Offered by Service Bureau." *Datamation* 14 (1968), 69.

[Malacara-Hernandez 02] D. Malacara-Hernandez. *Color Vision and Colorimetry: Theory and Ap-plications*. Bellingham, WA: SPIE Press, 2002.

[Newell et al. 72a] Martin E. Newell, R. G. Newell, and T. L. Sancha. "A New Approach to the Shaded Picture Problem." In *Proc. ACM Nat. Conf., p.* 443, 1972.

[Newell et al. 72b] Martin E. Newell, R. G. Newell, and T. L. Sancha. "A Solution to the Hidden Surface Problem." In *Proceedings of the ACM annual conference - Volume 1, ACM'72*, pp. 443–450. New York: ACM, 1972. Available online (http://doi.acm.org/10.1145/800193.569954).

[Phong 75] B. T. Phong. "Illumination for Computer Generated Pictures." *Communications of the ACM* 18:6 (1975), 311–317.

[Polyanin and Manzhirov 98] A. D. Polyanin and A. V. Manzhirov. *Handbook of Integral Equations*. Boca Raton, FL: CRC Press, 1998.

[Requicha 80] A. A. G. Requicha. "Representations for Rigid Solids: Theory, Methods, and Sys-tems." *ACM Computing Surveys* 12 (1980), 437–464.

[Roth 82] S. D. Roth. "Ray Casting for Modelling Solids." *Comput. Graphics and Image Process. (USA)* 18 (1982), 109–144.

[Shumacker et al. 69] R. A. Shumacker, R. Brand, M. Gilliland, and W. Sharp. "Study for Apply-ing Computer-Generated Images to Visual Simulation." Report AFHRL-TR-69-14, U.S. Air Force Human Resources Lab., 1969.

[Smith and Lyons 96] Alvy Ray Smith and Eric Ray Lyons. "HWB: A More Intuitive Hue-Based Color Model." *Journal of Graphics Tools* 1:1 (1996), 3–17.

[Smith 81] Alvy Ray Smith. "Color Tutorial Notes." Technical Report No. 37 37,

Lucasfilm, 1981.

[Strichartz 94] R. Strichartz. *A Guide to Distribution Theory an Fourier Transforms*. Boca Raton, FL:CRC Press, 1994.

[Sutherland and Hodgman 74] Ivan Sutherland and GaryW. Hodgman. "Reentrant Polygon Clip-ping." *Communications of the ACM* 17 (1974), 32–42.

[Sutherland et al. 74] I. E. Sutherland, R. F. Sproull, and R. A. Shumacker. "A Characterization of Ten Hidden Surface Algorithms." *ACM Computing Surveys* 6 (1974), 1–55.

[Torrance and Sparrow 76] K. Torrance and E. Sparrow. "Theory for Off-Specular Reflection from Roughened Surfaces." *J. Optical Soc. America* 57 (1976), 1105–1114.

[Warnock 69a] J. Warnock. "A Hidden-Surface Algorithm for Computer Generated Half-Tone Pictures." Technical Report TR 4–15, NTIS AD-733 671, University of Utah, Computer Science Department, 1969.

[Warnock 69b] John Warnock. "A Hidden Surface Algorithm for Computer Generated Halftone Pictures." Ph.D. thesis, University of Utah, 1969.

[Watkins 70] G. S. Watkins. "A Real-Time Visible Surface Algorithm." Report UTEC-CS-70-101, Dept. Comput. Sci., Univ. Utah, Salt Lake City, UT, 1970.

[Weiler and Atherton 77] K. Weiler and K. Atherton. "Hidden Surface Removal Using Polygon Area Sorting." *Computer Graphics (SIGGRAPH'77 Proceedings)* 11:2 (1977), 214–222.